MICROSCOPY
AND
PHOTOMICROGRAPHY
A WORKING MANUAL

MICROSCOPY AND PHOTOMICROGRAPHY
A WORKING MANUAL

Robert F. Smith, D.sc., DIP./R.M.S.,
R.B.P., F.B.P.A., F.R.A.S., F.R.M.S.
Emeritus Director of Biomedical Communications
New York State College of Veterinary Medicine
Cornell University
Ithaca, New York

CRC Press
Boca Raton   Ann Arbor   London   Tokyo

Library of Congress Cataloging-in-Publication Data

Smith, Robert F. (Robert Frank), 1917-
    Microscopy and photomicroscopy : a working manual / Robert F. Smith.
        p.   cm.
    Includes bibliographical references.
    ISBN 0-8493-8803-1
    1. Compound microscope--Handbooks, manuals, etc.
2. Photomicrography--Handbooks, manuals, etc.   I. Title.
    [DNLM: 1. Microscopy--methods.   2. Photomicrography--methods. QH 207 S658m]
    QH212.C6S55   1990
    502'.8'2--dc20
    DNLM/DLC
    for Library of Congress   90-10852
                                                                                                          CIP

Developer: Telford Press

This book represents information obtained from authentic and highly regarded sources. Reprinted material is quoted with permission, and sources are indicated. A wide variety of references are listed. Every reasonable effort has been made to give reliable data and information, but the authors and the publisher cannot assume responsibility for the validity of all materials or for the consequences of their use.

Direct all inquiries to CRC Press, Inc., 2000 Corporate Blvd., N.W., Boca Raton, Florida, 33431.

©1990 by CRC Press, Inc.

International Standard Book Number 0-8493-8803-1

Library of Congress Card Number 90-10852

Printed in the United States of America   3 4 5 6 7 8 9 0

Printed on acid-free paper

CONTENTS

FOREWORD

The microscope is one of the most versatile instruments available to science. One would be hard pressed to find a field of science or industry where the microscope is not used. However, it is also the most misused, abused and misunderstood of the precision instruments, and for this reason the author has attempted to compile a simple and practical working manual for those who use a microscope in their daily routine.

Specifically, this manual is intended for use by medical technologists, histologists, pathologists and students in the life sciences. The books contents eliminate all the complicated mathematics usually associated with microscopy and provides practical working information that permits users to quickly choose proper film for photomicrography, troubleshoot optical difficulties, and generally obtain maximum performance and image quality from their optical system.

The author wishes to acknowledge the following individuals for their cooperation and dedicated help in the preparation of this book. Mr. Lee Shuett, Nikon Inc.; Mr Ernst Keller, Carl Zeiss Inc.; Mr Mortimer Abramowitz, Olympus Corp.; Mr. Martin Scott, Eastman Kodak Co.; Mr. Robert Leonard, American Volpi Corp.; Mrs. Sandra Berry, director of Bio-Medical Communications New York State College of Veterinary Medicine at Cornell and her wonderful staff; John Lauber; Peter Daly; Tony DeCamillo; Mrs. Vivian Lawson; Deborah Dell; and Jane Jorgensen for her splendid artwork and color plate layouts.

Last but not least, I am grateful to my wife Jacqueline whose encouragement and support for the past fifty-five years has made possible everything worthwhile I have accomplished.

To all, my love and thanks.

Robert F. Smith

CHAPTER 1

BRIEF HISTORICAL BACKGROUND AND BASIC PRINCIPLES OF THE MICROSCOPE

In 1590 Hans and Zacharias Janssen of Middleburg Holland constructed the first compound microscope, i.e., a tube with a lens on either end, one of which could be placed close to the object (objective) and the other close to the eye (ocular). This crude but workable instrument introduced the basic concept of the modern microscope.

Antoni Van Leeuwenhoek used the simple strong magnifier and specimen holder in the year 1673 and observed infusoria at magnifications of about 275X. He discovered microorganisms in water and in 1683 published the first drawings of bacteria.

Giovanni Battista Amici, a university professor at Modena and Florence, became famous for his high quality microscopes and in 1827 introduced an instrument with matched achromatic systems. He later pointed out the importance of cover glass thickness to image quality and succeeded in improving both image quality and brightness by filling the space between the objective and the cover glass with water (water immersion objective).

The work of Amici was picked up and advanced by the famous team of Carl Zeiss, a master instrument maker, and professor Ernst Abbe, a physicist. Abbe improved on the Amici immersion system by using a suitable oil that possessed the same refractive index as glass. This is the oil immersion system commonly used today. Still not satisfied with the quality and color correction of microscope objectives, Zeiss and Abbe were joined by Dr. Otto Schott, a glass chemist. Schott was able to formulate glass melts that satisfied Abbe's calculations for color corrected objectives and in 1886 they announced the first apochromatic objectives for the microscope.

Early in the twentieth century, Professor Köhler introduced the method of microscope illumination which bears his name and is universally used throughout the world. He also designed and perfected a microscope for ultra-violet light. The design of this instrument was based on the principle that the shorter the wavelength of the light used for illumination of the specimen the greater the resolution possible with any given combination of optics. Abbe's classic experiments on the interference principle of image formation in the microscope and the importance of a high numerical aperture are demonstrated in a later chapter. Zeiss, Abbe, Schott and Köhler are what the author calls the Four Horsemen of Microscopy, as they are responsible for the development of the microscope as we know it today.

In order to secure a firm foundation in microscopy, it is necessary that we understand the exact purpose of the microscope. In most cases it will be stated that its prime purpose is to magnify. This is only partly true. The real, and most important, function is to resolve. Magnification in itself is not sufficient. High orders of amplification may be obtained without revealing fine structures. Magnification with the optical instrument is confined to very stringent limits and a good rule is to never exceed 1,000 times the numerical aperture of the objective. This will be taken up in detail later in the book.

How does a microscope magnify? The closer an object is brought to the eye the larger it becomes and the more detail we see. However, there is a limit, as close focusing of the human eye is limited. Normally 250mm is considered to be normal viewing distance at which an object is seen at 1X magnification. With young people this distance is somewhat shorter and focusing is possible at 125mm. This makes possible a 2X magnification, due to the fact that the image of the object occupies a larger area on the retina.

In order to obtain a closer look at objects, it is necessary to spread the image on the retina artificially. To spread this image, a magnifier must be used. The very strong magnifiers are called simple microscopes. They are capable of magnifications in the range of 250X. Such was the microscope used by Leeuwenhoek.

The modern microscope magnifies through two separate lens systems (after Hans and Zacharias Janssen), the eyepiece and the objective, and is, therefore, called a compound microscope. The slide projector projects a magnified image of the photograph on a screen. The magnification produced is a function of the focal length of the lens, and the distance of projection. If a 1m wide image of a 35mm slide (24×36) is projected, the magnification equals

$$M = \frac{1000\,mm}{36\,mm} = 28$$

Thus we see the projected image 28X larger than we would at 250mm. If this same projected image is viewed through a 10X magnifier the total magnification would be $28 \times 10 = 280X$. However, the field of view would be reduced by the 10X factor.

Microscope magnification is brought about in the same manner, *i.e.*, the total magnification is equal to the product of the magnification of the objective and the magnification of the eyepiece.

HOW IS THE IMAGE FORMED?

The structural elements that the microscope is called upon to resolve differ only slightly in refractive index. Consequently, they exert only a negligible influence on the light they transmit. The light they transmit is limited to a single characteristic of light waves which cannot be detected by the eye. What is changed is the phase of momentary vibration state. It is for this reason that conventional brightfield illumination will not reveal brightness differences between the structural details in the specimen and its surroundings. The image will lack contrast and details will remain invisible. However, when the emerging waves from the various structures have acquired larger phase differences due to differences in refractive index, greater contrast will be produced. This manifests itself by an edge effect (diffraction, refraction, and reflection). Under these conditions the production of a high contrast image of large structures does not present a problem. It should be noted that this edge effect can be carried too far by lowering the numerical aperture excessively. This will be explained further, and illustrated, later in the book.

Small structural details can only be revealed by changing the phase image into an absorption image by means of staining. Techniques such as darkfield, phase contrast, or Nomarski differential interference contrast (DIC) are other methods for obtaining contrast. These methods take advantage of differences in optical density, refractive index, and phase differences produced in the specimen. Further explanations will be given as we progress further into the book and cover the various types of illumination systems available to us today. The authors own method of additive color differential optical staining will be covered in the chapter on contrast methods and special techniques. As previously stated, it is not the intent of this book to dwell on the physics of optics but, rather, to provide a sound practical working knowledge in the daily use of the microscope.

CHAPTER 2

SET-UP AND ALIGNMENT

In order to be proficient in the use of the microscope, it is necessary to start with a sound foundation and understanding of the parts of the microscope and their function. In this book, we will progress step by step through all aspects of the handling of the instrument so that, by the time the reader has finished, all doubts and misconceptions should be dispelled.

First, let us take an overall look at the instrument. Figures 1 and 2 show a front and side view of a modern microscope equipped with an automatic camera for photomicrography. The essential parts have been labeled for quick reference.

There are two parts of the microscope that seem to be confusing; they are misunderstood, and the cause of much frustration and inferior results. These are the radiant field diaphragm, and the aperture diaphragm. Some refer to the radiant field diaphragm as just the field diaphragm. However I prefer to use the term radiant because that immediately associates it with the light source. The radiant field diaphragm is located at the light source and the aperture diaphragm is located in the condenser. The latter should never be used to control the intensity of the light as this will produce diffraction and destroy resolution. The intensity is controlled by either a rheostat or neutral density filters. Both of these diaphragms must be used properly in order to obtain the optimum Köhler illumination which is dealt with in detail later in this chapter.

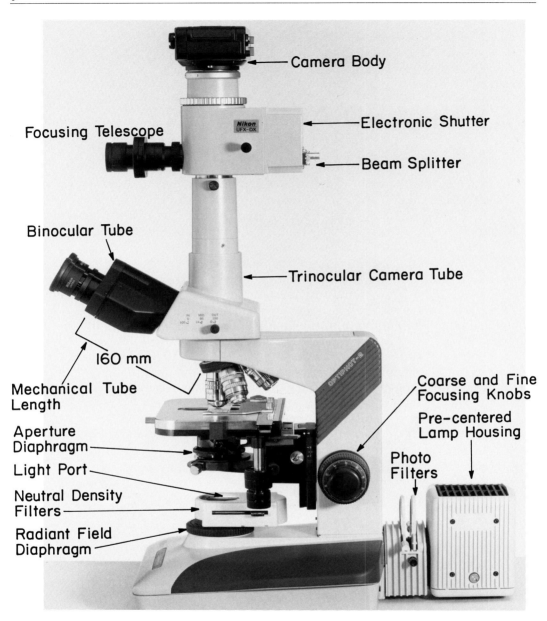

Figure 1

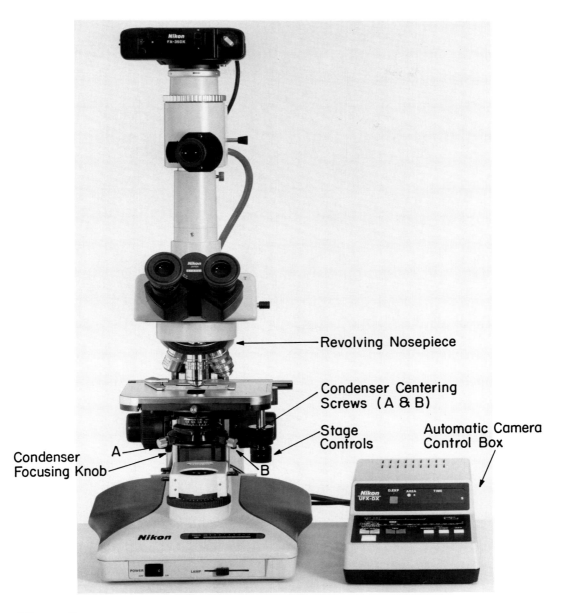

Revolving Nosepiece

Condenser Centering Screws (A & B)

Stage Controls

Automatic Camera Control Box

Condenser Focusing Knob

A

B

Figure 2

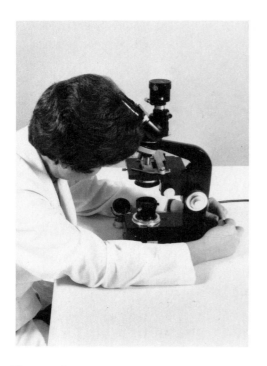

Figure 3

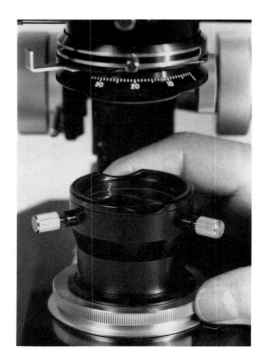

Figure 4

SETTING-UP THE MICROSCOPE

Alignment is the key word in microscopy. Without proper alignment there is no hope of obtaining satisfactory results. Therefore, we will start with a most basic and important element, the light source. To do this properly, either a target that replaces the condenser or a mirror may be used. As this varies with the make and model of the microscope, both procedures will be described and illustrated.

Figure 5

THE MIRROR METHOD

Using this method, a plano mirror is placed in such a position under the stage and to one side so as to reflect an image of the iris diaphragm of the condenser, as shown in Figure 3. The aperture diaphragm of the condenser is then closed as far as it will go. This will now serve as a projection screen on which to image the spiral filament of the lamp. First making certain that the radiant field diaphragm is wide open (Figure 4), turn on the lamp and observe if the filament is sharply defined and centered as in Figure 5.

Figure 6

Figure 6 shows the relationship of the mirror to the microscope as it should be. In Figure 7, A and B are the centering screws that are used to center the image of the filament on the iris of the condenser. At this point the filament may not be in sharp focus but this is easily remedied by turning the knurled ring C (Figure 7). This will permit the lamp socket to be moved in and out of the receptacle, and, by viewing the image of the filament in the mirror, it is easy to determine when the image is at its sharpest point. When this condition has been reached tighten the ring to secure the socket in its proper position. The centering and focusing procedure using the mirror system is now complete.

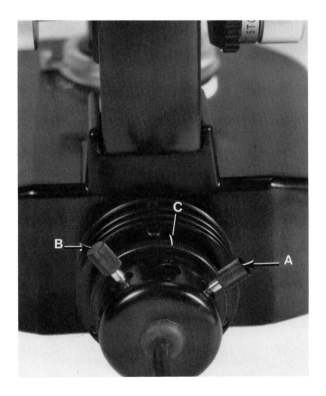

Figure 7

Figure 8

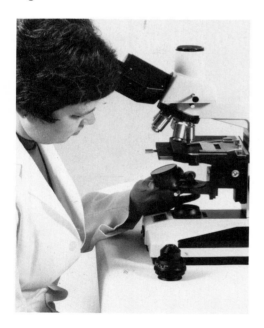

Figure 9

THE TARGET METHOD

This method is similar to the mirror method but is a little more convenient. In this method a plastic target (Figure 8) is used to replace the substage condenser. This plastic replacement contains a central area with a cross and two circles. The procedure is as follows: rack down the condenser and replace it with the target as shown in Figure 9. When this has been accomplished, rack up the condenser mount to its highest position. At this point one should be certain that the radiant field diaphragm is opened to its widest position. When the target is racked up to its proper position, an image of the filament will appear on the target area. If the filament is not in sharp focus and is off center, the two adjusting screws on the lamp

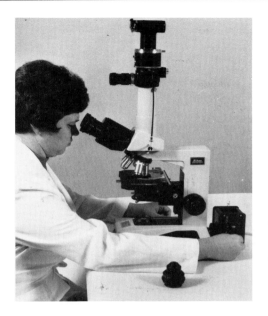

Figure 10

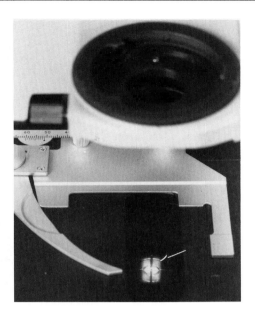

Figure 11

housing should be manipulated as shown in Figure 10 until an image appears, as shown in Figure 11. When using either the mirror or target method be certain that the diffuser has been removed from the light bulb so that a sharp image of the spiral filament may be obtained. The binocular housing has been removed in Figure 11 in order to obtain a better camera angle of the target. Figure 12 is a closeup of the lamp housing showing the controls A and B for focusing and centering the filament on the target. The lamp housing also has provision for inserting contrast or color correction filters, as shown in C.

Figure 12

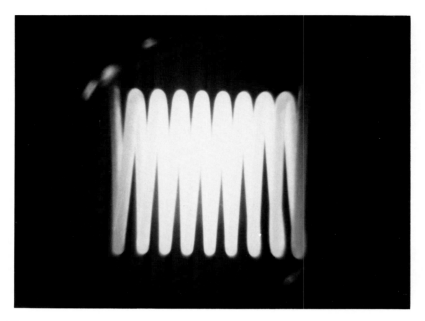

Figure 13

The sole purpose of this procedure is to ensure even illumination. The filament is focused at the lower plane of the condenser and not at the specimen plane. The filament may be observed as explained in the two foregoing procedures or by removing an ocular and looking down the tube to observe the rear focal plane of the objective (Figure 13).

The image will not be seen by looking through the oculars. Most microscopes are equipped with auxiliary diffusion discs to render the spiral filament invisible when using low power lenses. However, the absence of a diffuser does not have a detrimental effect on the image at high magnification, as can be seen in Figure 14. The absence of the diffuser actually improves the contrast. The filament is not visible because, having been focussed at the plane of the condenser aperture diaphragm, it does not come to focus in the same plane as the specimen, but only at the rear focal plane of the objective. Had the filament been improperly focused, it would appear in focus with the specimen, resulting in dark bars across the field, as shown in Figure 15.

Having completed the first steps as described, we may now proceed to the next procedure necessary for obtaining Köhler illumination and to proper adjustment of the instrument for comfortable viewing over long periods of time without fatigue or eye strain.

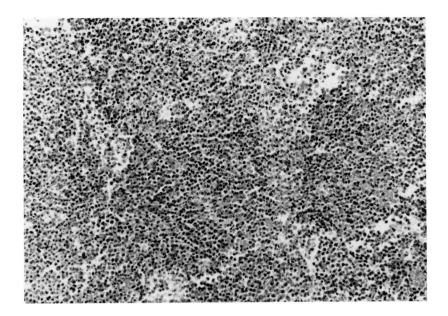

Figure 14

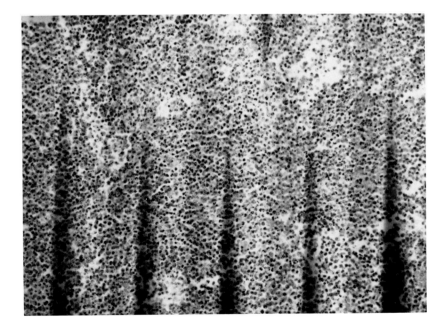

Figure 15

Figure 16

Because the spacing between the eyes varies from one individual to the next, it is necessary to adjust the oculars to suit the individual observer. To do this, place a slide on the stage, turn on the lamp, focus the specimen, and adjust the inter-pupilary distance by bringing the oculars closer together or further apart as shown in Figure 16. When a single field is seen, the images of the right and left eyes have become fused and observation may be carried out in complete comfort. To readjust to this position without wasting time, simply use the scale shown in Figure 17 (arrow). This setting should be remembered or recorded, and should it be changed for any reason, it is a simple matter to reset it to the proper number for your interpupillary distance.

Figure 17

Figure 18

The next step is the adjustment of the oculars. This is very important as it is possible to see the image sharply with one eye and slightly out of focus with the other. Some fortunate individuals have normal vision (emmetropia) while others may be nearsighted (myopia) or farsighted (hyperopia). If the two latter conditions are mild they can be compensated for by a slight focusing adjustment of the microscope. However, if the refraction problem is such that corrective lenses must be worn in order to perform normal duties, then glasses should always be worn when using the microscope.

Figure 19

The first step in adjusting the oculars is to determine which eye is dominant. To do this, form a mask with both hands and frame a distant object. The object should be viewed with both eyes open. Then while viewing, and without moving the head or hands, close one eye and then the other. When doing this you will notice that the object will disappear out of the frame with one of the eyes. The eye that keeps the object framed is the dominant eye. When the dominant eye has been determined, the next step is to have both oculars adjusted to the zero diopter setting. The diopter settings on some oculars is displayed by a numerically calibrated scale, but on most of the latest instruments the zero setting is represented by a white line circumventing the ocular (Figure 18). Initailly, both oculars should be set at the white zero line. In addition to the diopter settings each ocular is equipped with eye-cups (Figure 19). The eye-cups should be extended for non-spectacle wearers and retracted when spectacles are worn. The purpose of this is to keep the eye the proper distance from the ocular, so that the entire field may be observed.

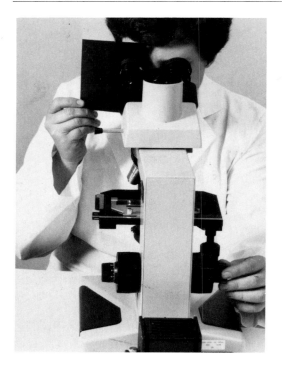

FINAL ADJUSTMENT OF THE OCULARS

With both oculars set at zero, place a specimen slide on the stage and bring the image into the best possible focus. Now place a black card in front of the non-dominant eye as shown in Figure 20. In this case the operator's dominant eye is the left. If the opposite were the case the sequence would be reversed. Using the dominant eye, focus the specimen as critically as possible using the coarse and fine adjustments, as shown in Figure 21.

Figure 20

Figure 21

When the image is at its sharpest point, transfer the card to the other eye (Figure 22). At this point the coarse and fine adjustments must not be changed from the position obtained with the dominant eye. Now view the image with the non-dominant eye and bring it into sharp focus by adjusting the ocular collar, as shown in Figure 22. A closeup of the ocular adjustment is shown in Figure 23. The oculars are now adjusted for each eye, ensuring viewing comfort and no eye strain. This procedure must be repeated each time the microscope is used if more than one person is using the instrument.

Figure 23

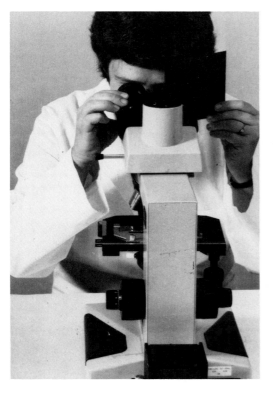

Figure 22

Figure 24

KÖHLER ILLUMINATION

Up to this point, we have concentrated on the centering and focusing
of the light source and adjustment of
the oculars. Now we are ready to align
the optical components of the microscope. To do this, the two diaphragms
mentioned earlier as being vital to good
microscopy will be adjusted and
centered to provide the ultimate in
image quality. To begin, both the
radiant field and the aperture
diaphragms will be opened to their
maximum aperture.

It is necessary here to point out that
the controls for the radiant field
diaphragm may be situated differently
in various microscope models. However, regardless of how or where they
are situated, they perform precisely the
same function. The radiant field
diaphragm is not the only adjustment

Figure 25

that can be found in different locations. Figures 24, 25, and 26 show the sub-stage assembly of two different models of Nikon microscopes. Figure 24 shows the radiant field diaphragm adjustment directly behind the lightport. This is a departure from some microscopes, where this adjustment is sometimes located on the side of the base or on the lightport itself. Figure 25 demonstrates the location of the adjustments on the same model microscope as seen in Figure 24. A and B are the condenser centering screws. The use of these adjustments will be described later in this chapter. In the same Figure, C is the condenser aperture adjusting ring and D releases the stage tension so that it may be rotated. This is a very useful feature in photomicrography. In Figure 26 an entirely different approach to adjustment is illustrated. Instead of the centering screws being located on the condenser carrier, they are situated on the lightport collector lens. A and B are used for centering, while C is a combination diffuser for very low power objectives and an additional two positions for the 40X and 100X objectives. D is the radiant field diaphragm. E is the condenser aperture control and F is a lever for displacing the condenser aperture to obtain oblique illumination.

Figure 26

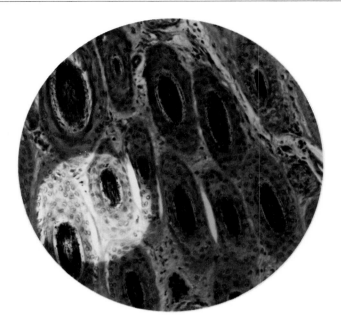

Figure 27

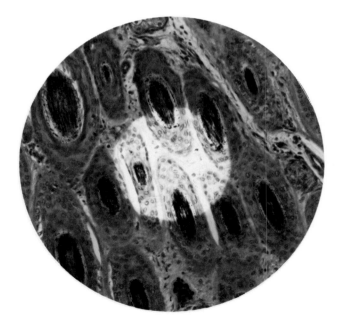

Figure 28

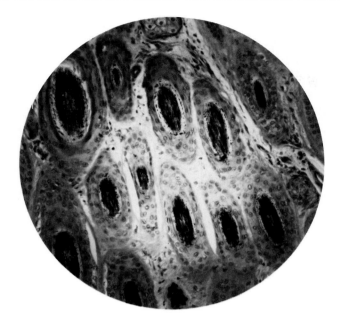

Figure 29

With the controls described, it is now possible to make the final adjustments for Köhler illumination. The first step is to place a slide on the stage and rotate the 10X objective into place. Turn the lamp up to a comfortable brightness and focus the specimen. At this point the radiant field diaphragm and the aperture diaphragm should be at their widest openings. Now stop the radiant field diaphragm to a small opening. If it appears as shown in Figure 27, the condenser is badly off center. By using the condenser centering screws as shown in Figures 25 and 26, the spot of light may be brought to the center of the field, as shown in Figure 28. Notice that the leaves of the diaphragm and the specimen are both in sharp focus. If the diaphragm appears as in Figure 29 it means that the condenser is centered but improperly focused. To remedy this, the condenser must be focused.

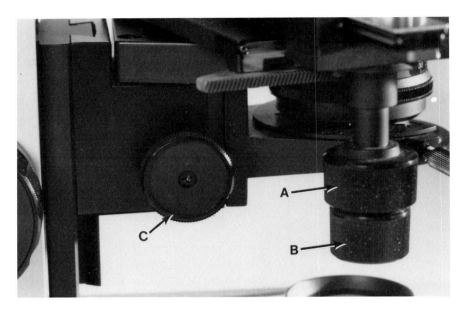

Figure 30

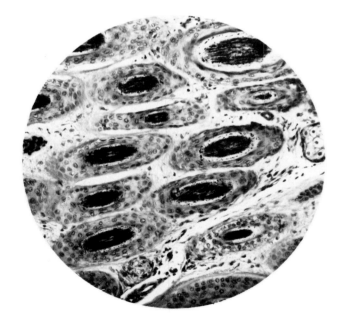

Figure 31

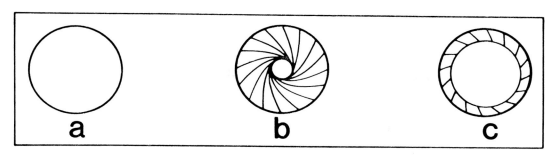

Figure 32

Figure 30 shows some of the substage controls. A and B are the X and Y stage movement controls and C is the condenser focusing knob. By focusing the condenser up or down you will arrive at an image similar to that in Figure 28. At this point the radiant diaphragm should be opened until it just disappears from the field of view, as shown in Figure 31.

The next step requires adjustment of the aperture diaphragm. The diagram in Figure 32 shows three settings of the aperture diaphragm. Remember, the aperture diaphragm is situated in the condenser. It is extremely important that this diaphragm be adjusted in such a way as to be compatible with the total numerical aperture of the optical system and the nature of the specimen.

The nature of the specimen is an all important factor and it determines the optimum setting of the aperture diaphragm in the condenser. Specimens differ widely in their optical characteristics and therefore the aperture settings shown in Figure 32 are "rule of thumb" settings. With most stained sections it is a safe rule to follow. However, if the specimen was a diatom or a blood smear, conditions would dictate a deviation from the ideal setting (C) in Figure 32. The aperture setting may be observed by removing an ocular and looking down the tube to the rear focal plane of the objective. A phase centering telescope will permit a larger and better view of the aperture.

In the following chapter the terms refraction, diffraction absorption and dispersion will be used; therefore, an explanation of these phenomena is in order.

Absorption

Many microscopic specimens are stained to reveal their structures. When light passes through them its intensity is reduced selectively depending on color and density. This selective absorption makes possible the revelation of fine structures in the specimen. The selective absorption of the wavelengths comprising white light produces colored light. An optical staining technique taking advantage of this phenomena will be described later in this book.

Refraction

The change in direction of a ray of light passing from one transparent medium into another of different optical density *i.e.* from air to glass. A ray passing from a less to a more dense medium is bent in a direction perpendicular to the surface. The amount of deviation depends on the wavelength. The shorter the wavelength the greater the deviation.

Diffraction

The bending of light rays around objects with sharp edges. When a wave front meets a sharp edge it generates a new wavefront at that edge. A slit aperture in an optical system generates a diffraction pattern. At small apertures the definition is lowered. This is demonstrated later, in Chapter 3.

Dispersion

The separation of a beam of light into its constituent wavelengths as a result of refraction on entering a transparent medium. The change of refractive index with wavelength. A classic example is the spectrum produced by a prism, or in nature a rainbow.

NUMERICAL APERTURE

In the previous chapter three settings of the aperture diaphragm were shown (Figure 32). The correct setting of the diaphragm is most important as it has a major influence on the total numerical aperture of the system. However, as already stated, the proper setting of this aperture will vary with the nature of the specimen.

Each objective has engraved on its barrel the type of objective, *i.e. Apochromat, Achromat, etc.*, the magnification, numerical aperture (N.A.), tube length and the coverglass thickness for which it is corrected. A typical example would be PLAN-APO-20X 0.65 N.A. 160/0.17. At this time we will deal with only one of these markings, the N.A. or numerical aperture. It is this factor that determines the optical resolution of your system.

Resolving power is directly related to numerical aperture: The higher the numerical aperture, the greater the resolution. Resolving power is the property by which an objective shows two small elements in the structure of an object which are a short distance apart to be distinctly separated. The measure of resolving power is the N.A. The higher the N.A. the greater the resolving power of the objective and the finer the detail it can reveal. Numerical aperture is given by the formula N.A. $= n \sin u$, where n is the lowest refractive index that appears between the object and the front lens of the objective, and u is half the angular aperture of the objective. In other words the more indirect or refracted and diffracted the rays that an objective is capable of accepting, the greater its revolving power. When the condenser and its numerical aperture, which is controlled by the aperture diaphragm, closely match that of the objective the full potential of the objectives resolving power may be realized. The influence that variations in these settings have on the image is demonstrated in the following groups of photomicrographs.

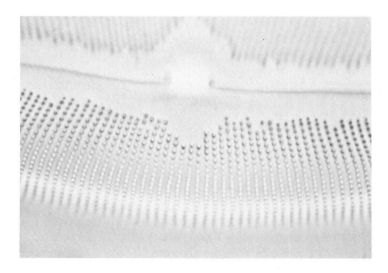

Figure 33A

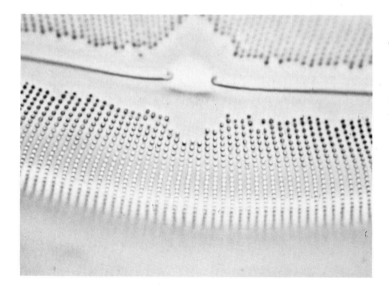

Figure 33B

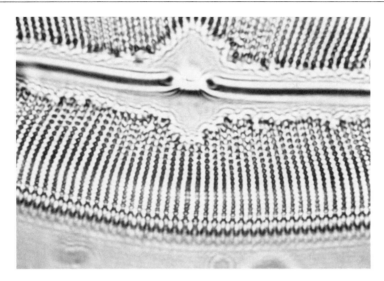

Figure 33C

Figure 33 A-B-C. Specimen, Diatom; Magnification, 1,500X. Objective 100X Plan Apochromat N.A. 1.35. Condenser - Achromatic Aplanatic N.A. 1.40 oiled to the slide. Specimen type - Diffractive and refractive absorption extremely low.

Figure 33A. Aperture diaphragm matching the objective N.A. 1.35 90%. Notice the lack of contrast and low visibility of specimen details such as the punctae.

Figure 33B. Aperture diaphragm at 50%. Here the contrast has increased and the resolution is excellent showing distinct separation between the punctae.

Figure 33C. Now if we carry the stopping down to excess with the diaphragm stopped down to 12% or less the punctae are no longer well defined and in fact run together, forming lines instead of rows of dots. This is excessive diffraction and more examples will follow using Abbe's theory of image formation.

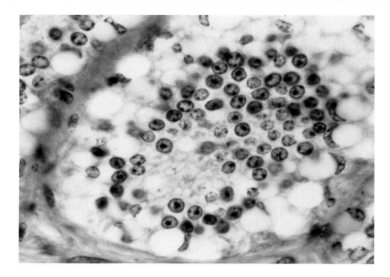

Figure 34A

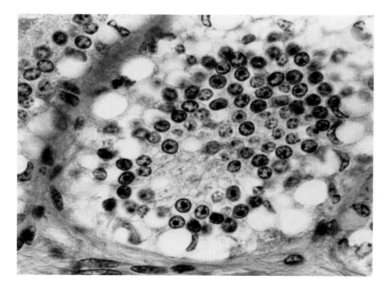

Figure 34B

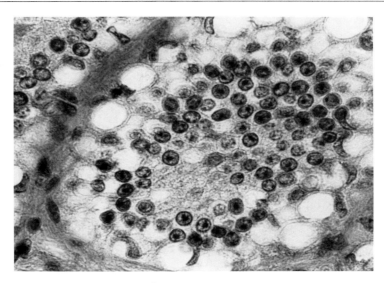

Figure 34C

Figure 34-A-B-C. Specimen, Dog kidney stained section; Magnification, 500X. Objective 40X Plan-Apo. N.A. 0.95. Condenser Achromatic Aplanatic N.A. 1.40 oiled to slide.

Figure 34A. The colors in a stained section demonstrate strong absorption and very little diffraction and refraction quality. The important details are revealed at a high setting of 90%.

Figure 34B. At a 50% setting maximum detail and contrast are obtained and for such a specimen this is the ideal setting.

Figure 34C. With the aperture diaphragm set at 12% or less most of the important details are lost due to the broadening of fine lines as a result of diffraction and refraction.

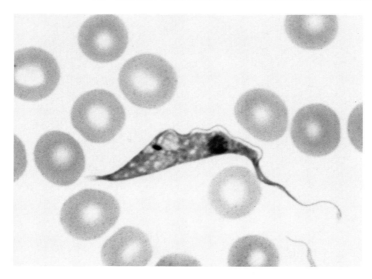

Figure 35A

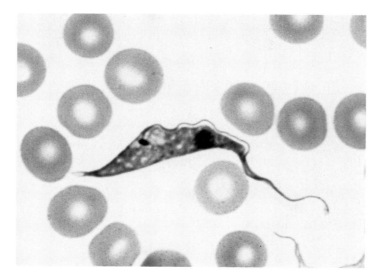

Figure 35B

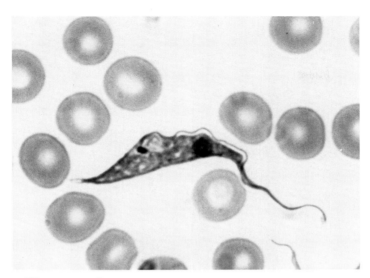

Figure 35C

Figure 35-A-B-C. Specimen, *Trypanasoma lewisi* in rat blood; Magnification, 1,500X. Objective Plan-Apo. 100X N.A. 1.35. Condenser Aplanatic Achromatic N.A. 1.40 oiled to slide.

Figure 35A. With a specimen such as this very little contrast is gained by the reduction of the aperture diaphragm, as there are minimal edge effects due to diffraction and refraction. This photomicrograph was made at an aperture setting of 90%.

Figure 35B. Here the aperture was reduced to 50% and only a very slight contrast gain is noticeable.

Figure 35C. Here the aperture is reduced to less than 12%, but there is only a slight indication of diffraction around the red blood cells and the undulating membrane of the organism.

Figure 36

Figure 37

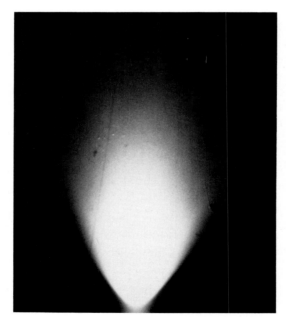

Figure 38

IMPORTANCE OF N.A.

If a very narrow central pencil of rays is used for illumination, the finest detail that can be shown by a microscope with high enough magnification is equal to *w.l./N.A.*, where *w.l.* is the wavelength of the light used for illumination. The wider the pencil used for illumination, the greater the resolving power up to a maximum at which the width of the pencil is sufficient to fill the whole aperture of the, objective. Under these circumstances the resolving power is twice as great and the finest detail the objective can now show is equal to w.l./2 N.A.

This same limit is reached when a narrow pencil of the greatest possible obliquity is used. For example, the wavelength of the brightest part of the spectrum may be assumed to equal 0.00053mm. Therefore, an objective of N.A. equal to 1.00 will resolve two lines separated by a distance of

$$\frac{0.00053}{1.00} = 0.00053 \, mm$$

with a narrow central illuminating cone and

$$\frac{0.00053}{2 \times 1.00} = 0.000265 \, mm$$

with a cone filling the whole aperture or with a narrow oblique cone. A 40X, 0.85 N.A. objective should theoretically resolve lines separated by distances ranging between 0.00062 and 0.00031 mm depending upon the mode of illumination. For a 40X, 0.65 N.A. objective the limiting values are 0.00081mm and 0.000405mm.

It is obvious from the forgoing that the ability of an objective to resolve is dependent upon the N.A.. The higher the N.A. the greater the angle of the cone of illumination it will accept and which will fill its entire aperture. It is also evident that the objective N.A. is limited by the N.A. of the condenser. This does not mean that using a high N.A. objective with a lower N.A. condenser is not beneficial because, as stated in chapter 2, the more refracted and diffracted light from the specimen the objective is capable of collecting, the better the resolution. This is demonstrated further along in the chapter.

Let us now take a look at the angle of the cone of light emitted by the condenser when the aperture diaphragm is stopped down, fully opened, and in oil contact with the slide. Figure 36 demonstrates the narrow cone of light produced when the aperture diaphragm is stopped down to its smallest opening. Such a cone of light will produce an image that has great contrast and very low resolution due to diffraction. Figure 37 shows the cone of light produced by a condenser of N.A. 1.40 when the aperture is fully opened but is not in oil contact with the slide. Figure 38 demonstrates the increased angle of illumination when the same N.A. 1.40 condenser is placed in oil contact with the slide. This cone of light will fill the back lens of a 1.35 or 1.40 objective and therefore utilize the full optical potential of the lens. An important point must be made here. If the same 1.40 condenser were to be used with a 10X N.A., 0.4 a 20X, 0.65 or even a 40X of 0.85 or 0.95, there is no benefit in having the condenser in oil contact with the slide. With the 10X 0.4, 20X 0.65 and 40X 0.85 or 0.95, the N.A. of the dry condenser will be 0.95 and this exceeds the cone of acceptance of these objectives and will require stopping down the diaphragm in order not to flood the objective with excess light resulting in very poor contrast. With the 40X 0.95 objective the N. A. will be matched but will still require some reduction in the aperture diaphragm in order to obtain optimum image quality. Naturally, just how far the aperture is reduced will depend largely on the nature of the specimen, as previously explained and demonstrated.

Ernst Abbe, in his quest for a fully corrected high resolution objective (the apochromat) which he perfected in 1886, asked himself two questions: First, why does a large aperture produce more perfect images and finer detail than a small one, when in practice the incident cone of rays may only fill a small aperture while the remaining apparently unused portion remains as a dark space? Second, what is the mechanism that makes the use of this dark space possible? The dark space is shown in the diagram in Figure 39. This represents two settings of the aperture diaphragm and is a view of the rear focal plane of the objective. On the left the aperture is set at about 90% and on the right at 12% or less. In spite of the fact that in neither case is the full aperture of the objective filled with direct rays, a high aperture objective will capture the diffracted rays from the dark area and produce a higher resolution image than an objective with a low numerical aperture.

After countless experiments investigating the dark space Abbe came to the conclusion that the secret lay in diffraction at the object structure, and therefore could only be explained in terms of the wave nature of light. To confirm this concept he devised a whole system of experiments that remain to this day the most beautiful and impressive example of the physical experimenter's art. These experiments make it possible for the nonexpert to understand Abbe's diffraction theory of image formation in the microscope. Some of these experiments have been duplicated here. A regular divided grating will substitute for a specimen because it will most graphically illustrate that all objects for microscopy have the structure of an absorption or phase grating. Every specimen is, in the optical sense, an irregular grating consisting of many very small areas of different optical properties; absorption, density, diffraction, refractive index, *etc.*

However, before we start with the diffraction apparatus and the many experiments it allows, a brief introduction to Abbe's theory is in order. Abbe concluded that geometric optics alone could not explain resolution or the process of image formation. Only the wave theory of light is able to satisfy that condition.

A point is due to the wave nature of light imaged as a diffraction disc. G.B. Airy divided specimens into two categories, Selbstleuchter (producing their own light) and Nicht-Selbstleuchter (not producing their own light) in the, elements of the image. Two points can be resolved if the light intensity between them is at least 80% of that of the main maxima (each in the center where the image of the point is supposed to be).

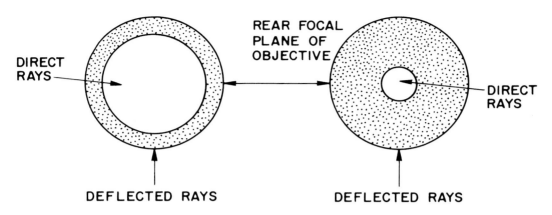

Figure 39

Illuminated points of a specimen emit coherent waves because they are hit by parallel wave fronts of an infinitely distant light source. These coherent waves interfere and produce an interference figure. Amplitudes of the single elements of the latter have to be summed. Their square values give the light intensity.

Airy's method did not explain all the factors on which the resolution depends. The solution to this problem is due principally to Ernst Abbe. He assumed that the, microscope objective images not only the specimen but also the light source illuminating the specimen. He assumed that the microscope objective images not only the specimen but also the light source illuminating the specimen. Provided the light source is very small and at infinity (auxiliary lenses) a single-plane elementary wave travels through the specimen and is made convergent by the objective. In the rear focal plane of the objective, one finds the image of the light source (the so called primary image). From this point the wave continues to travel, however now as a divergent spherical wave, and at a certain distance behind the objective it produces a homogeneously illuminated area. A specimen brought into the beam will influence and change the image of the light source, which will then reproduce the specimen.

The following is a description of one of Abbe's experiments that appears in PLATE I in the color section of this book. The first object in the Abbe diffraction plate is a line grating, Figure 1A, with spacings of 16 microns on top and 8 microns on the bottom. All lines are resolved because all diffraction maxima are present, as shown in Figure 1B, the primary or diffraction spectrum of the object. In all cases, this diffraction spectrum can be observed only by removing the ocular and looking down the tube, or by using a phase centering telescope. Here, we see the main maxima (center) and the neighboring 1st, 2nd, and 3rd orders.

We will now demonstrate the progressive deterioration of resolution as these orders are removed by reducing the N.A. of the objective. Figure 2A shows the diffraction spectrum with just the main maxima and two neighboring maxima on either side. When this condition exists, the resulting image of the grating has lost most of its resolution, as shown in Figure 2B. The lines on the top are still resolved but the finer bottom lines have now all but disappeared.

The results of progressing further with aperture reduction is shown in Figure 3A. Here only the main maxima is present. Figure 3B shows the resulting image of the grid. No lines are resolved. A single diffraction maximum, regardless of what order it is, cannot produce an image that is similar to the specimen.

The line grating is now replaced by a point grating, Figure 4A. The distance between points is 12 microns, and all are resolved due to all the orders being present, as shown in Figure 4B. If a 1mm slit is placed in the system, all orders except those in the north-south direction are eliminated, Figure 5A under these conditions, the dots of the grid now become lines, Figure 5B.

If the slit is rotated, as shown in Figure 6A, the line system rotates simultaneously. The lines are always perpendicular to the slit Figure 6B. The dots are only resolved in the direction of the diffraction spectrum. When the interference orders are not present, the dots are not resolved and fuse to become lines. Many of the interference orders are present in the dark area previously described and it is for this reason that the greatest resolution is only obtained with high N.A. objectives regardless of the condenser used. However, maximum image quality is only obtained when the N.A. of the condenser nearly matches that of the objective. When an oil immersion objective with an N.A. greater than 1.00 is used, the condenser should be oiled to the slide.

All plates appear following page 135.

CHAPTER 4

CONDENSING SYSTEMS AND THEIR USE

THE CONDENSER

This most important component of the microscope is often misused and misunderstood. It is the determining factor in the production of high resolution images, *i.e.*, images that will be compatible with the potential resolution of the objective being used.

Figure 40

There are basically five different types of condensers used in microscopy: Abbe, achromatic-aplanatic, darkfield, phase contrast, and Nomarski differential interference. Three are shown in Figure 40. From left to right, they are aplanatic-achromatic, Abbe, and darkfield. In most laboratories the Abbe is the hands down favorite. It is well suited for routine work and its flip out top element makes switching from low to medium power very simple. However, for research and color photomicrography the choice would be the achromatic-aplanatic condenser, because it is free of aberrations and its color correction closely matches that of the finest objectives.

That the numerical aperture be large is as important in a condenser as in an objective. The numerical aperture of an Abbe condenser may be 0.95 but its aplanatic cone does not exceed 0.45. At this point an explanation of the difference between total numerical aperture and the size of the aplanatic cone is in order, as it is on the latter that the total efficiency of the microscope depends. Figure 41a shows that, with a simple uncorrected lens, the marginal rays do not come to focus at the same point as the central rays. The lens is not corrected for spherical aberration. This diagram is exaggerated. There even is a portion of a simple lens that will nearly bring the incident rays to a common focal point. However, it is

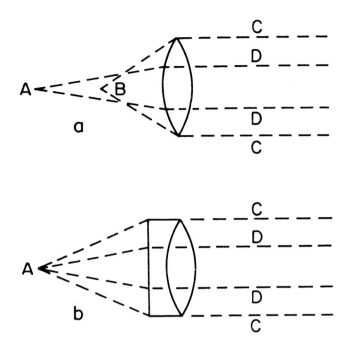

Figure 41

obvious that the central portion is small in relation to the whole. Therefore, the aplanatic cone, which is the largest cone of light that the lens will bring to a single focal point with a small or point light source, is in this case very small. It may be assumed that the aplanatic portion of this condensing system is represented by the incident beam included between DD, which focuses at A, whereas the outer portion between D and C will focus progressively nearer the lens depending on its distance from the center. The peripheral zone C will therefore be focused at B and the remainder between A and B.

Figure 41b shows a lens system corrected for spherical aberration. The size of the aplanatic cone now approaches the total aperture of the condensing system and the marginal rays are focused either exactly or very nearly in the same position as the central rays. The degree to which this can be accomplished depends on the size of the aplanatic cone.

Light Cone   **Image Produced**

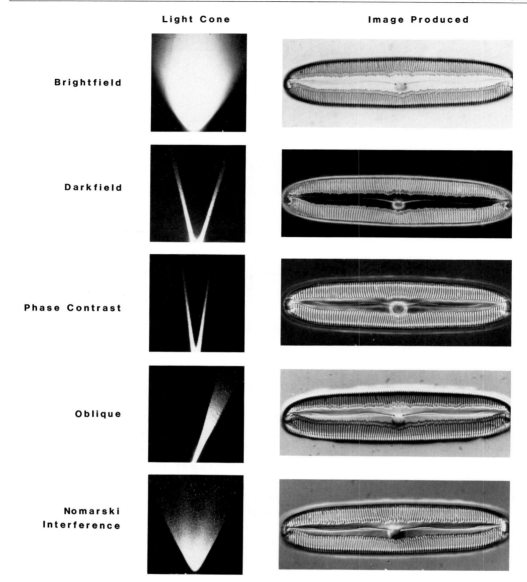

Brightfield

Darkfield

Phase Contrast

Oblique

**Nomarski
Interference**

Figure 42

To obtain a better understanding of the condenser cone of light and the resulting image, reference should be made to Figure 42. This plate illustrates the different types of illumination available to the microscopist for specific specimen characteristics. From top to bottom, we have brightfield, darkfield, phase contrast, oblique illumination, and Nomarski differential interference. Notice the similarity between the oblique illumination image and that of Nomarski interference. The former mode is a much neglected type of illumination and should be explored by the microscopist when possible. It can, when finances prohibit the acquisition of more sophisticated equipment, be used as a substitute with great success.

Oblique illumination can be acheived by decentering the aperture of the condenser (not the condenser itself). However, since not many condensers are supplied with this feature today, other methods must be sought. Actually, this can be quite simple if one has a combination brightfield phase contrast condenser. The procedure is as follows: Obtain the normal brightfield image and proceed to rotate the condenser ring toward the first phase disc. In between the two the aperture diaphragm will be displaced laterally to a sufficient degree to produce an oblique cone of light. It will be very obvious when the correct degree of obliquity has been reached since the image will have a strong relief appearance. This can be seen in Figure 42.

In addition to oblique illumination, darkfield is also possible using the combination condenser. This can be accomplished only with objective magnifications up to 16X, but it serves as a most useful tool for many specimens. To obtain the darkfield image, the 100X phase ring is used in combination with a 4X, 10X or a 16X objective. Best results are obtained with the 10X objective. If higher magnifications in darkfield are required, a proper darkfield condenser is necessary.

If budget permits, it is possible to have all the conditions shown in Figure 42 with a single condenser. Such condensers are supplied by all the manufacturers for their laboratory and research microscopes.

The last condition, Nomarski interference, shows the cone of light to be similar to brightfield but, if closer examination is made, it will be noticed that the cone of light is composed of light and dark areas. These areas are the interference bands produced by the Wollaston prism in the condenser.

When deciding which optical equipment to purchase, it is wise to select components which can provide all modes of illumination. No one system replaces another. On the contrary, they complement each other.

CORRECTION		
Achromats	**Flourite**	**Apochromats**
2 colors chromatic aberration 1 color spherical	2 colors chromatic 2 color spherical	3 colors chromatic 2 color spherical

Figure 43

OCULAR TYPE		
Achromats	**Flourite**	**Apochromats**
100X = Compensating 40X = Compensating 20X = Compensating 10X = Huygenian 4X = Huygenian 2X = Huygenian	100X = Compensating 40X = Compensating 20X = Compensating 10X = Compensating 4X = Compensating or Huygenian 2X = Compensating or Huygenian	100X = Compensating 40X = Compensating 20X = Compensating 10X = Compensating 4X = Compensating 2X = Compensating

Figure 44

CHAPTER 5

OBJECTIVES

Microscope objectives for general microscopy are of three different types, *achromatic, fluorite,* and *apochromatic.* These objectives have not changed materially in their degree of correction since the nineteenth century. While in recent years they have been computed to produce flat fields, they are still dependent upon a special type of ocular. The objective is not completely corrected in itself. It has an error which must be compensated for in order to eliminate a difference in lateral chromatic aberration. To do this, and because it is compensating for an error in the objective, the term "compensating ocular" is used. Figure 43 shows the degree of correction for each type of objective and Figure 44 pairs the proper ocular type with each category of objective.

Recently, a new generation of optics has been introduced by Nikon Inc. Known as CF optics, these are so finely corrected that the achromats rival the more highly corrected objectives of former years. While there are many complicated physical differences between the CF and conventional compensating systems, it is not our purpose to discuss them here. What is important to note is that in the CF system one type of ocular serves both the Achromat and Apochromat.

The Achromatic objective is really the prototype of the modern microscope objective. It was developed by Lister and Amici in the early nineteenth century and efforts were made to correct spherical and axial chromatic aberration simultaneously. To better understand the various corrections in objectives and oculars, we will take a look at the basic principles involved.

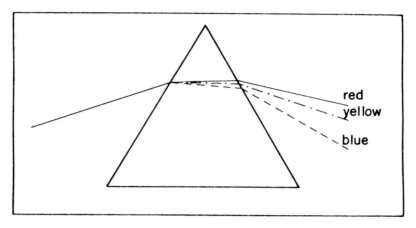

Figure 45

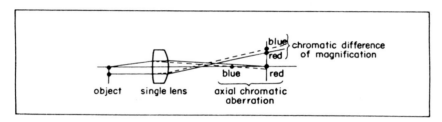

Figure 46

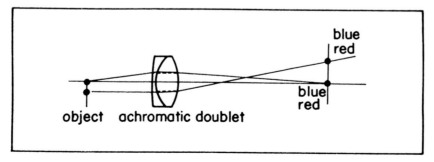

Figure 47

Due to the fact that the refractive index of the material used for an optical system depends on the wavelength of the light passing through it, it is possible to observe the phenomenon called dispersion. We have all observed in our high school science classes white light passing through a prism and spreading out into a spectrum (Figure 45). Since every optical material has its own dispersion factor, optical glasses are classified into flint glass, which has higher dispersion, and crown glass, which has lower dispersion. If an optical system is made up of a single convex lens of crown or flint glass only, it does not form a sharp image because of dispersion, as shown in Figure 46. This lens would be classified as having severe chromatic aberration. The horizontal distance between the axial image is called axial or longitudinal chromatic aberration, while the vertical difference in height is called lateral chromatic aberration or chromatic difference in magnification. The accepted way to correct chromatic aberration is by combining a convex lens of crown glass with a concave lens of flint glass. By combining the two properly, chromatic aberration is corrected, as shown in Figure 47. In the CF system the chromatic difference in magnification does not have to be corrected in the eyepiece, therefore, the axial chromatic aberration can be corrected more easily and accurately. Both objective and eyepiece are complete in themselves in the CF System. Chromatic aberration manifests itself by the appearance of color fringes around very fine structures.

Figure 48

The correction made by the manufacturer of microscope optics can all be for naught if the equipment is not used properly. A case in point is the use of coverslips of the wrong thickness. Microscope objectives are of two types; those corrected for use with and without coverslips. Those corrected for use with coverslips should be used with a No. 1½ thickness, not 1, 2 or 0. The No. 1½ ranges in thickness from 0.16mm* to 0.19mm. This is within the tolerances for which all medium to high power objectives are corrected. The coverslip thickness for which an objective is corrected is marked on the mount along with the N.A., the magnification, and tube length, as shown in Figure 48. The objective on the left is corrected for no coverslip and the other is marked 0.17, which is a No. 1½. Contrary to popular belief, the oil immersion objectives are not-immune to the aberration caused by the use of coverslips of incorrect thickness or uncovered specimens. Such use results in a somewhat hazy, less sharp, image and, while not as severe as with the medium and high dry lenses, the full potential of the objective can not be realized when coverslips vary from the No. 1½ thickness. An example of this image deterioration may be seen in Figures 49 and 50. In Figure 49 the specimen was not covered but photographed with an oil immersion objective corrected for specimens covered by a No. 1½ coverslip. Compare this with Figure 50, which was photographed with an objective corrected for no coverslip.

* Most range from 0.17 to 0.19mm.

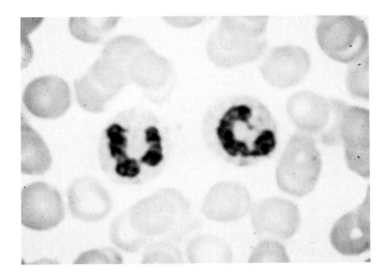

Figure 49

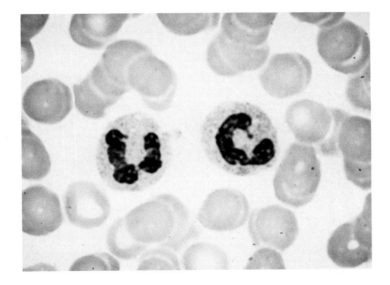

Figure 50

Peduncity.

O blood supply on.

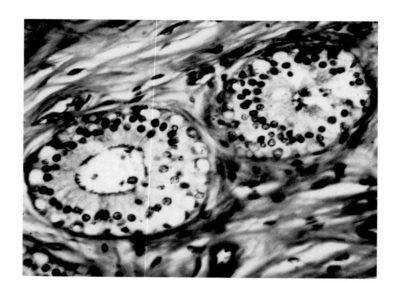

Figure 51

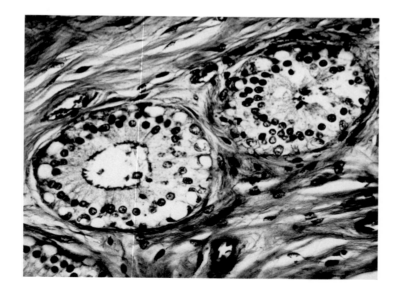

Figure 52

When possible, oil immersion should always be used for medium to high magnifications since the numerical aperture is much higher and, therefore as previously explained, the resolution will be directly proportional to the N.A. An example of this may be seen in Figures 51 and 52. The 100X photomicrograph in Figure 51 was made with a 10X objective N.A. 0.4 and a 10X ocular. Compare this with the 100X photomicrograph in Figure 52 which was made with a 40X N.A. 0.95 objective and a 2.5X ocular. This demonstrates the importance of using the highest N.A. available.

The manufacturers supply immersion oil to be used with their objectives. This oil has been selected not only for its refractive index of 1.515 but also for a dispersion factor compatible with their optical design. Remember, the N.A. of an oil immersion objective is only valid when the condenser is in oil contact with the slide. This forms what is called a homogeneous oil immersion system.

CORRECTION
COLLAR

Figure 53

When a homogeneous oil immersion system is not possible and a 40X high dry objective must be used, be certain to use one with a correction collar as shown in Figure 53. This will make possible correction for any deviation from the 0.17mm coverglass thickness. Should the coverglass be too thick or too thin when using the high dry objective the results can be disastrous as shown in Figure 54. However, by using the correction collar all spherical aberration may be eliminated as shown in Figure 55. To achieve this with the correction collar the image should be brought into the best possible focus using the fine focusing adjustment. Be certain that the condenser aperture diaphragm is fully open when making this adjustment. If there is any hazy appearance to the image, this means that the coverglass is either too thick or too thin. If this is the case, slowly turn the collar to the left while correcting focus with the fine adjustment. If the image does not improve, try the opposite direction, making certain to be constantly correcting with the fine focus. When the image becomes sharp and haze free, continue to turn the collar until the image begins to haze over again. At this point turn back until the sharpest image is obtained. Now stop down the condenser aperture diaphragm to its optimum setting and you have completed the correction for coverglass thickness. This exercise will have to be repeated for each slide if the optimum in image quality is to be obtained.

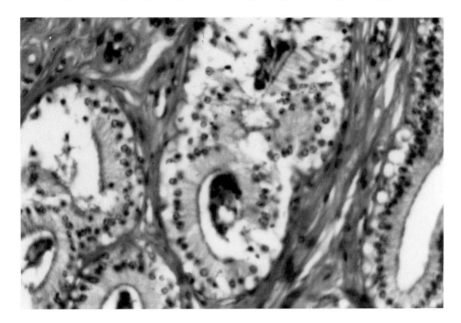

Figure 54

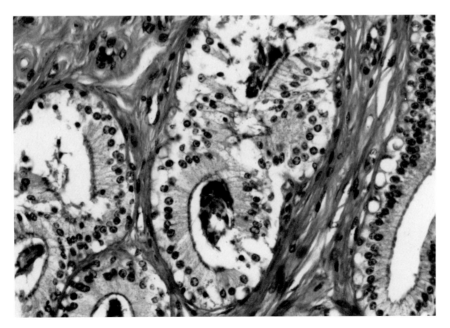

Figure 55

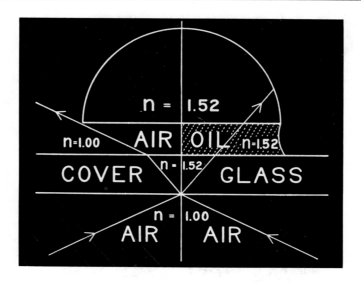

Figure 56

It is obvious now that the coverglass is an integral part of the optical system. In fact, it is the first lens of the objective. Coverglasses are plain glass plates; however, plain glass plates have definite optical properties and therefore have been computed into the design of the objective. With very low power, low numerical aperture objectives of 0.03 or less, specimens may be observed and photographed covered or uncovered without noticeable spherical aberration. With higher N.As. the optimum 0.17mm coverglass should be used for best results. Figure 56 illustrates the advantage of the oil immersion objective over a dry system.

It should be noted here that not all objectives are corrected for a tube length of 160mm. American Optical has long made use of infinity corrected objectives and just recently Zeiss introduced their color free ICS (infinity color corrected system). These optics generate a space of parallel light bundles into which all components for contrast-enhancing techniques, including phase contrast, DIC and polarization, can be incorporated without altering the performance of the optical system. There are many different opinions regarding the two systems and it has to be left up to the individual to make a choice.

TABLE I. CF EYEPIECES			
Type	**Magnification**	**Field Number**	**Focal Length (mm)**
High eyepoint ultrawide field	CFUW 10X	26.5	25
	CFUW 10XM*	26.5	25
High eyepoint wide field	CFW 10X	18	25
	CFW 10 XM*	18	25
	CFW 15X	14	16.7
Photo eyepieces **	CF photo 2X	21.6	78.5
	CF photo 2.5X	18	63.5
	CF photo 4X	11.8	35
	CF photo 5X	9.4	28.5

* With picture frames
** Exclusively for photography

OCULAR FIELD OF VIEW

When using a particular objective ocular combination it is often desirable to know the size of the field being observed. All oculars have a field of view number. Table I presents the type, magnification, field of view number and focal length of various types of CF oculars. Conventional compensating oculars also have similar field numbers. These factors may be obtained from your dealer or manufacturer. To determine the diameter of the field being observed, the field of view number is divided by the prime magnification of the objective. For example, a CFW 10X ocular with a field number of 18 used with a 10X objective would display a field of 1.8mm. An ocular with a field of view number of 26.5 used with the same objective would display a field of 2.65mm, *etc.* Most modern oculars have larger field of view numbers than those of past years.

CHAPTER 6

PHASE CONTRAST AND NOMARSKI INTERFERENCE

We have discussed the basic systems of illumination for unstained and stained specimens, but there are specialized methods such as phase contrast and Nomarski interference which offer many advantages to the serious microscopist studying unstained living material. Here we will deal first with the most widely used routine method namely, phase contrast illumination. Phase contrast reveals details in specimens possessing very slight differences in optical path or refractive index from the surrounding medium.

This system was invented by Zernike in 1932 and won him the Nobel Prize in physics. In a phase contrast system the phase of the central beam is changed by one quarter of a wavelength at the rear focal plane of the objective. This converts the imaging conditions of transparent objects into absorbing objects. Basically, it consists of an annular stop in the substage condenser for furnishing an annular ring of light in the rear focal plane of the objective. This separates the central rays from the diffracted rays. The objectives have a phase retardation plate built into the rear focal plane to retard the central beam by a quarter of a wave and a layer of metal deposited on the ring to absorb 75% of the passing light. Depending upon the initial magnification, this diffraction plate is placed between, or on the back lenses of, the objective.

Specimens of varying thickness and refractive index act as diffraction or phase gratings and result in unaltered central rays and deviated diffracted rays. The diffracted rays are brought to focus at the ocular where they reinforce the central rays, producing a bright image. The phase plate reduces the intensity of the central beam and results in a contrasting dark back ground for the reinforced bright image. This is called bright contrast and is one of several types of phase optics available.

Phase objectives are of the plano or flat field type and come in either achromat, fluorite, or apochromats. Naturally, the apochromats produce the most aberration-free images but, although preferred, they are not essential for excellent results. Because we are dealing only with a black and white image, a narrow band 550 nm green filter may be used with achromats, thereby eliminating troublesome aberrations that would be encountered with white light. The next consideration is the condenser and, as with the objectives, the choice and remedy are the same. However, as the condenser is a combination bright field phase contrast and will be

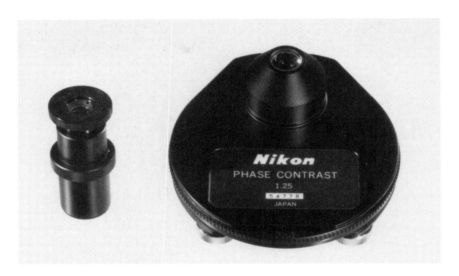

Figure 57

used for other types of microscopy, the more highly corrected aplanatic-achromatic type is desirable. Figure 57 shows a typical combination condenser with its focusing telescope.

In setting up for phase contrast observation, start off with the phase condenser in its bright field position. A stained specimen should then be placed on the stage and the condenser centered by checking the radiant field diaphragm for Köhler illumination. Next, an unstained specimen is placed on the stage and a low power phase objective is chosen and the appropriate phase ring in the condenser is brought into position and checked for centration and adjusted if necessary.

The next step is to remove one of the oculars and replace it with the focusing telescope shown in Figure 57. Looking through this centering ocular will reveal the phase rings of the system. If they do not appear in sharp focus, turn the top ring of the telescope until the phase plate appears in sharp focus. Figure 58 shows the phase rings in improper alignment. Should they appear like this, the resulting image will be unsatisfactory (Figure 59).

Figure 58

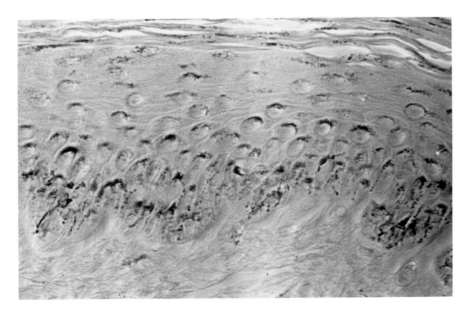

Figure 59

Figure 60

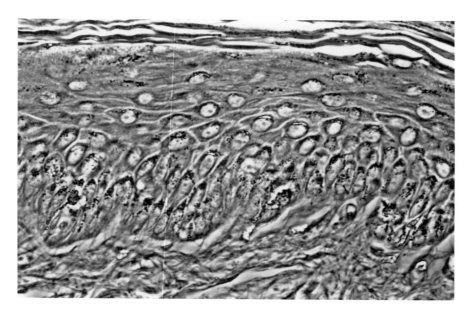

Figure 61

NOMARSKI		
Objective	Condenser	Shear in plane of object
16 × N.A. 0.32	Setting I	1.32 μm
40 × N.A. 0.65	Setting II	0.55 μm
100 × N.A. 1.25	Setting III	0.22 μm

Figure 62

By turning the phase ring centering adjustments on the condenser, the phase rings may be brought into perfect alignment (Figure 60). With the rings in proper relationship to each other, a strong contrast phase image will result as shown in Figure 61. Naturally, when the proper conditions have been satisfied the centering telescope is removed and the ocular replaced for normal observation. This procedure should be repeated with each change of magnification since two objectives seldom line up perfectly.

A word of advice is necessary at this point. Never use phase objectives for any serious brightfield observation or photomicrography. In spite of what salesmen might say, they just do not produce satisfactory brightfield images.

Nomarski interference is the most recent of the true interference contrast methods. Optically, it is far more complicated than the phase contrast system. Therefore, without getting too involved in the physical principles of the optics involved, the following explanation should help the reader have a better under standing of differential interference contrast microscopy.

Basically, the Nomarski interference system consists of a polarizer at the light port of the microscope and Wollaston prisms in the condenser. One prism for each magnification is arranged in a turret mount. Phase rings may also be provided so that the condenser can be used for brightfield, phase and Nomarski interference. Situated above the objectives is a second Wollaston prism and an analyzer. When the light passes through the polarizer and enters the first Wollaston prism, it is split into two plane-polarized components. One beam vibration is perpendicular to the plane of the prism while the second one is parallel to the plane. Both beams pass through the specimen in parallel and are separated by an extremely short distance (Figure 62). Above the specimen the two beams are recombined by the objective and the second Wollaston prism. The analyzer is oriented at an angle of 45° with respect to the vibration plane of each of the entering waves. The image formed by the existing rays is noticeably different from that of the phase system in that it possesses a strong relief, or pseudo three-dimensional, appearance. This pseudo three-dimensional image also creates the impression that surface details are being observed. This impression is false for most biological specimens. The three-dimensional or single azimuth shadow cast effect is the result of the angle of shear.

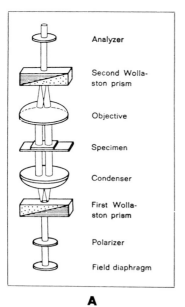

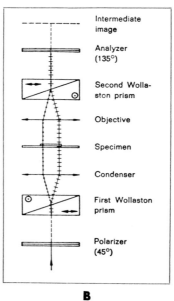

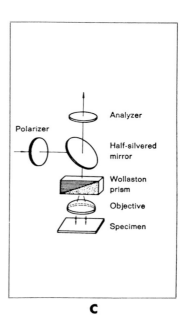

A

B

C

Figure 63

Figure 63 A shows the principles of beam splitting and recombination. Figure 63 B indicates the direction of vibration at various points in their path and only the axis of the light beams. The orientation of the polarizer is such that the natural light emerging from the radiant field diaphragm (not shown) is plane-polarized and has an inclination plane of 45° with respect to the plane of the diagram. In the lower part of the Wollaston prism a polarized wave entering it is split into two plane-polarized components. The dotted beam's vibration is perpendicular to the plane of the diagram while that of the cross-lined beam is parallel to the plane. Both beams pass through the specimen and are separated by an extremely short distance (Figure 62). Above the specimen the two beams are recombined by the objective and the second Wollaston prism. The analyzer is oriented at an angle of 45° with respect to the vibration plane of each of the entering waves. This insures equal intensity of the two beam components. In Figure 63 A and B it has been demonstrated that two Wollaston prisms are necessary for transmitted light. With Epi or reflected light as illustrated in Figure 63 C, only one prism is required. This is due to the fact that when light travels through the reflected light objective and the Wollaston prism it acts in the same manner as the condenser prism used for transmitted light. When the light is reflected back from the opaque specimen, it must travel through the Wollaston prism a second time, and this is, in effect, the same as the objective and second Wollaston prism used for transmitted light.

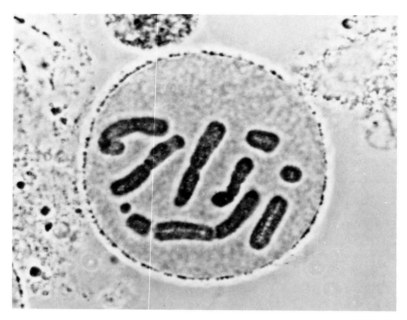

Figure 64

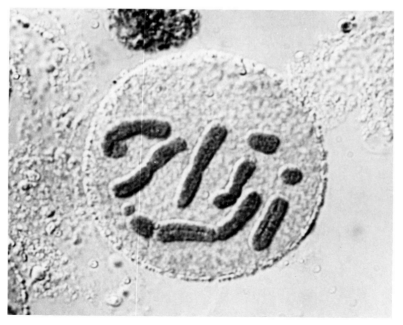

Figure 65

To summarize, the following points must be remembered. In phase microscopy, phase details are made visible due to differences in the refractive index or thickness of the specimen. When there are uniform phase details only areas containing gradients of steep refractive index will appear as different intensities in the image. Phase details are also made visible by the Nomarski system, but appear, as described earlier, as apparent relief or shadow cast images. The background image in Nomarski interference contrast is the area in the specimen, or, in the case of cells, the slide itself, where no object is present. With white light used to illuminate the object the background can be made to appear colored, black and white or grey. Therefore, color phenomena may be produced regardless of the presence of an object in the light path. This instrument characteristic affords a multitude of optical staining possibilities, some of which appear in Plate II in the color section of this book. For a direct comparison between a phase and a Nomarski image see Figures 64 and 65. The three dimensional effect and the illusion of depth and surface detail is very apparent in Figure 65.

All plates appear following page 135.

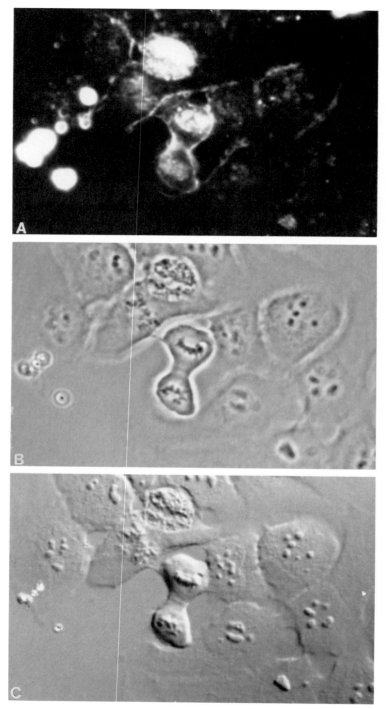

Figure 66

As stated in Chapter 4, no one system replaces another; rather they are complimentary and each yields its own type of information. The photomicrographs of HeLa cells in tissue culture in Figure 66 substantiates this fact. (A) darkfield, (B) phase contrast, and (C) Nomarski differential interference.

CHAPTER **7**

TROUBLESHOOTING

Regardless of how proficient one becomes in the use of the microscope, there is no immunity to image faults. At times, such as when dust collects on one of the surfaces of an optical component, it is both difficult and time consuming to remedy the trouble. In the case of dust and its removal, it often happens that we replace dust with new dust, or something worse. Therefore, a proper selection of solvents and tools is necessary. A good selection is as follows: Q-tips (wooden applicator sticks), lens paper, ether, Xylol, rubber ear syringe, small sable brush, and Kodak lens cleaner. To remove dust from a lens surface first degrease the sable brush by dipping it in a small quantity of ether (about 10cc) shake it dry and gently dust off the lens surface. Follow this by blowing off any remaining dust with the ear syringe. Do not use canned air to remove dust, as it can leave a residue that is harder to remove than the original problem. Finger prints can usually be removed with the Kodak lens cleaner by moistening a triple-folded piece of lens paper and gently rubbing the surface, followed by a rubbing with a triple-folded dry piece of lens paper and then removing any remaining lint with the syringe. To remove immersion oil, first wipe off the major portion with triple-folded lens paper. Follow this with a triple-folded piece of lens paper dampened with Xylol (do not saturate the paper). Wipe the lens dry and follow with lens paper dampened with Kodak lens cleaner and then finally polish dry with lens paper and again blow off any dust with the syringe. A word of caution here: be certain not to use the ether near open flames or cigarettes, as it is highly explosive. To store the ether, cap the can tightly and place it in a certified explosion proof refrigerator. If such storage is not available, omit the use of ether.

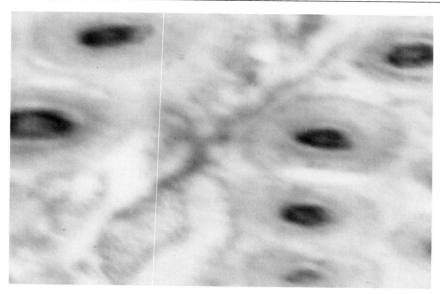

Figure 67

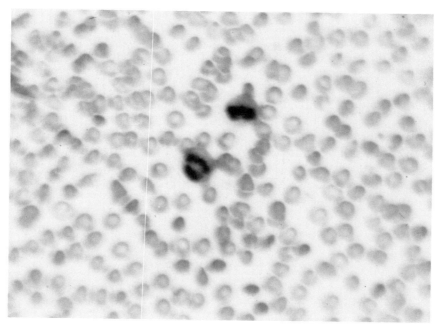

Figure 68

Figure 69

Figure 67 shows an image as it appears through a medium dry objective which has a smear of immersion oil on the front lens. With low, or high dry objectives, the front lens is the first place to look when this condition exists. However, in the case of a high dry objective, there is also the possibility of improper coverglass thickness, as described in Chapter 5.

Not all image faults are due to oil, dust, or fingerprints on the optical surface. Figure 68 shows an image formed by an oil immersion objective that is very hazy and blurred. If one removes an ocular and peers down the tube, he or she will more than likely see the culprit shown in Figure 69. A bubble in the oil is all it takes to destroy image quality and this is easily observed by removing the ocular. When this happens the bubble may be broken by swinging the objective left and right. Objectives should never be lowered into the oil. They should be swung in with a swiping motion since this avoids the trapping of air between the slide and front lens of the objective.

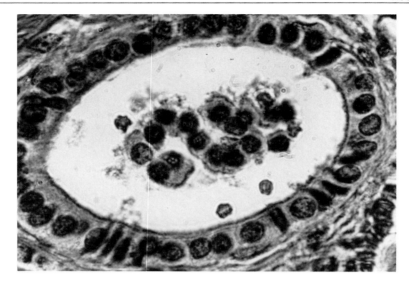

Figure 70

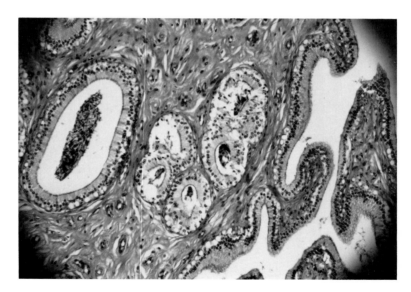

Figure 71

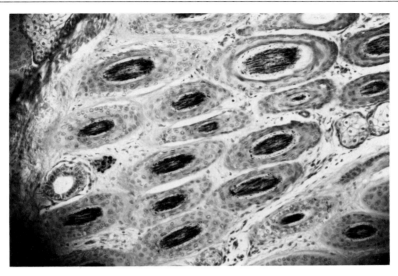

Figure 72

Another common fault is an image that appears overly contrasty, with halos surrounding fine structures, as shown in Figure 70. This condition is due to diffraction and is the result of excessive stopping down of the aperture diaphragm. The remedy, in this case, is to remove the ocular and observe the rear focal plane of the objective. Open up the condenser aperture diaphragm until it occupies ⅔ to ¾ of the total illuminated area of the back lens of the objective as described in Chapter 3.

It will be remembered that at the beginning of this book it was stated that two of the most troublesome components of the microscope are the aperture and radiant field diaphragms. We have just seen in Figure 70 what excessive stopping down of the aperture diaphragm can do and now Figure 71 demonstrates the effect of not opening up the radiant field diaphragm sufficiently. As prescribed in Köhler illumination, the radiant field diaphragm should be centered, focused, and opened up until it just disappears from the periphery of the field. This same or similar appearance of the field may result from the the use of a condenser of unsuitable focal length being used with a very low power objective. If the front lens does not flip out or an auxiliary lens is not provided, the only remedy is to change the condenser to one providing a larger area of illumination.

If the radiant field diaphragm is wide open and the image still appears as in Figure 72, it is certain that the microscope is badly misaligned. In that case, start from the beginning and follow procedures prescribed for Köhler illumination (Chapter 2).

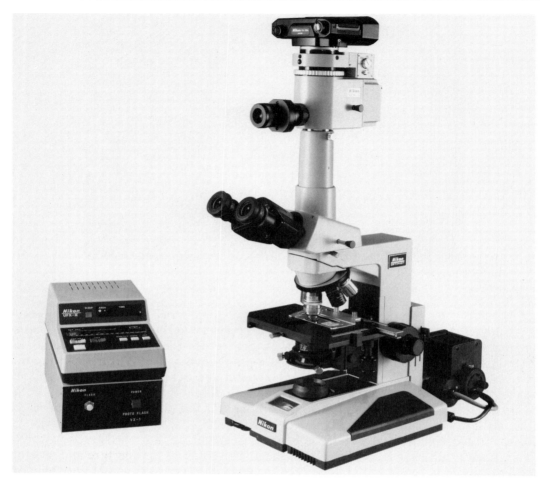

Figure 73 - Photo courtesy of Nikon Inc.

CHAPTER 8

PHOTOMICROGRAPHY

Photomicrography is the wedding of two disciplines, microscopy and photography. Today, one rarely visits a laboratory, research or industrial, where the microscope is not used. It is also rare not to find a camera of some sort attached to one or more of the microscopes in the laboratory. Photomicrography is used to document research, prepare teaching material and for publication. What the author hopes to present in this chapter is a sound introduction and practical working guide to this most important aspect of microscopy.

Before getting into the subject however, it might be of interest to the reader to become acquainted with some of the modern instruments designed for microscopy and photomicrography. They range from the simple and moderately priced, to the very sophisticated and expensive. All those shown perform well, but some excel. The choice should be made carefully and the author advises selecting a set of test slides and photographing identical fields with each instrument before making a purchase decision. This may seem rather involved, and time consuming, but it will avoid regrets in the long run.

Shown in Figure 73, is the Nikon OPTIPHOT microscope with attachment camera UFX. Stacked on the control panel is the new electronic flash power supply. This gives the user the choice of automatic continuous light photography or electronic when photographing highly motile specimens or suspensions. The UFX camera features both spot and integrated reading, plus a memory bank and multiple exposure capability. It may be equipped for all types of illumination including epi-fluorescence.

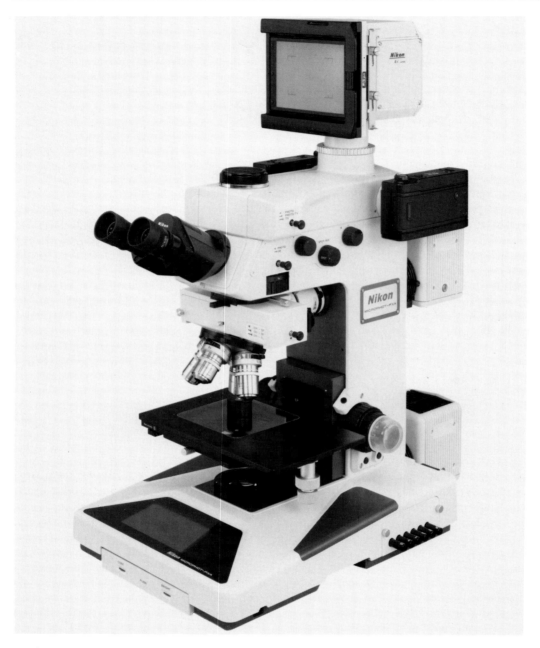

Figure 74 - Photo courtesy of Nikon Inc.

Figure 75 - Photo courtesy of Nikon Inc.

Figure 74 shows Nikon's top of the line MICROPHOT FXA. This instrument carries two automatic 35mm cameras plus a large format camera and a port for a TV camera. There is little this instrument will *not* do. It has a liquid crystal screen and a pull-out, IBM-compatible, computer keyboard in the base which has a built in intervalometer for time lapse studies plus complete data recording capability. For those having difficulties focusing low power photomicrographs there is automatic focusing with objectives 1X to 20X. For those using attachment cameras as shown in Figure 73, Nikon supplies a data recording camera back shown in Figure 75.

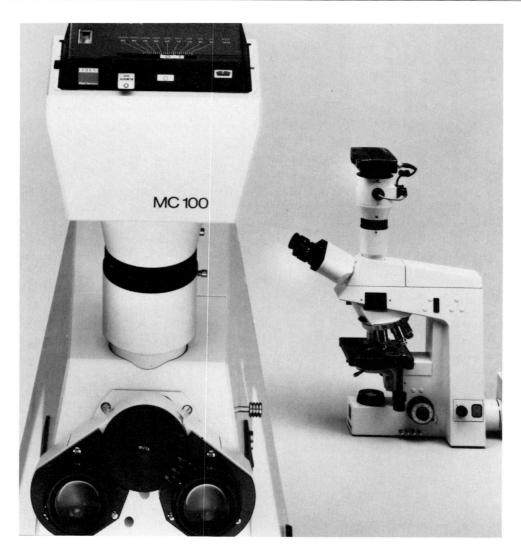

Figure 76 - Photo courtesy of Carl Zeiss Inc.

In Figure 76, the Zeiss AXIOSCOP microscope is shown with the NC 100 attachment camera. This a fully automatic camera and the microscope may be equipped with a choice of all the Zeiss optical systems and optics. Shown in Figure 77, is the top of the line Zeiss AXIOPHOT. This instrument features two 35mm automatic cameras, plus a 4 X 5 large format camera (not shown) and a port for a

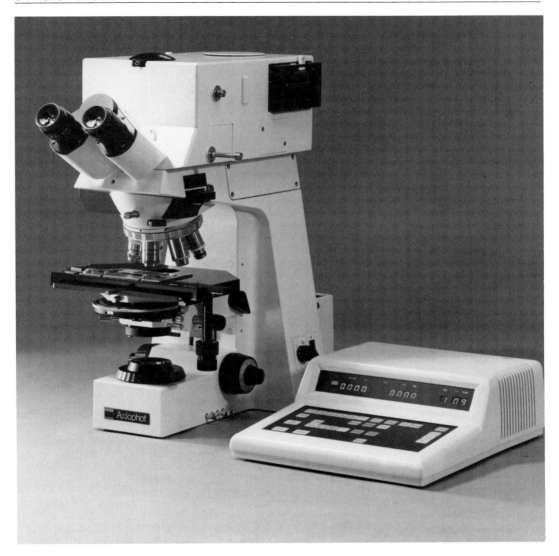

Figure 77 - Photo courtesy of Carl Zeiss Inc.

CC TV camera. It can be used for all microscopic techniques and features a photo multiplier sensor that displays a constant sensitivity over the entire spectrum.

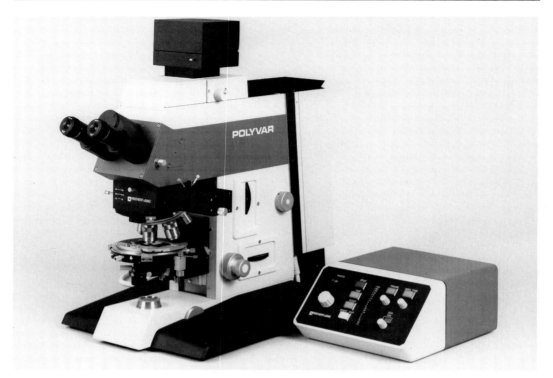

Figure 78 - Photo courtesy of Reichert Jung

The Reichert POLYVAR is a complete photomicrographic system that includes macro capability in both transmitted and reflected light. The instrument is shown with its single 35mm camera in Figure 78. An interchangeable large format camera is also available. Reichert also offers more modest instruments for routine laboratory applications.

Illustrated in Figure 79, is the Leitz DIALUX 20 microscope mounted with the VARIO-ORTHOMAT automatic camera. The camera permits spot and integrated readings to be made and may be mounted on most of the Leitz line of microscopes.

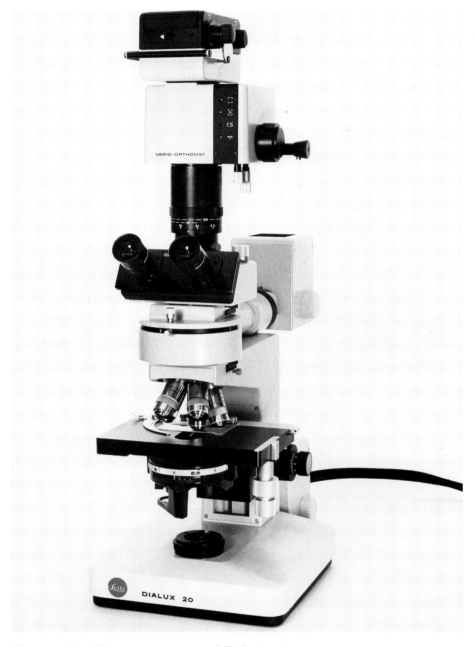

Figure 79 - Photo courtesy of E. Leitz Inc.

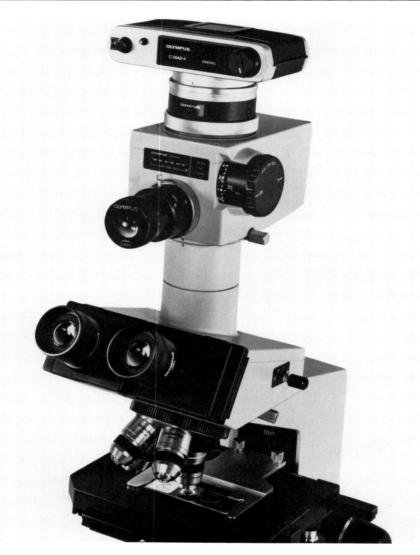

Figure 80 - Photo courtesy of Olympus Corp.

Olympus Corporation offers a complete line of photomicrographic systems ranging from the very modest PM-10AK shown in Figure 80, to the very elaborate. The PM-10AK is a semi-automatic system intended for the modest budget. In Figure 81, we see the top of the line AH-2. This microscope has two 35mm cameras mounted on both sides and a large format camera and TV port on the top. As shown here it has the 8 X 10 format mounted, but 4 X 5 is also available. It features spot

Figure 81 - Photo courtesy of Olympus Corp.

and integrated exposure readings, memory, motorized substage and iris and auto focus up to the 10X objective.

Regardless of how simple or complex the equipment, the principles of photomicrography are the same for all.

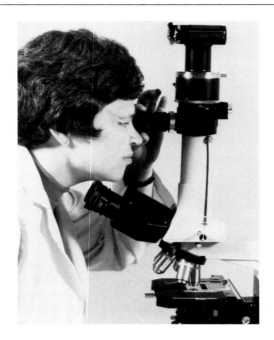

Figure 82

Figure 83

PHOTOMICROGRAPHY

Let's begin with the camera. We will limit the format size to 35mm, not because 35mm is best suited for all types of photomicrography, but because it is the one most commonly used. I should like to mention here that, for color 35mm is the practical choice. For black and white, however, there is no substitute for the 4 X 5 cut film format since each photograph may be custom made by using the type of film and processing most suitable for the specimen.

Most photomicrographic cameras today are automatic; consequently, exposure problems have been greatly reduced. However, picture sharpness seems to pose great problems for many who must record fields as a matter of routine. The first step is to parfocalize the camera with the format reticle. In Figure 82, the operator is turning the focusing telescope to bring the reticle into sharp focus. Figure 83 shows a closeup of the viewing and focusing telescope. The arrow points to the + and - calibrations which will indicate the proper setting for your eyesight. This reading may be recorded for future reference, thus avoiding the necessity of refocusing each time.

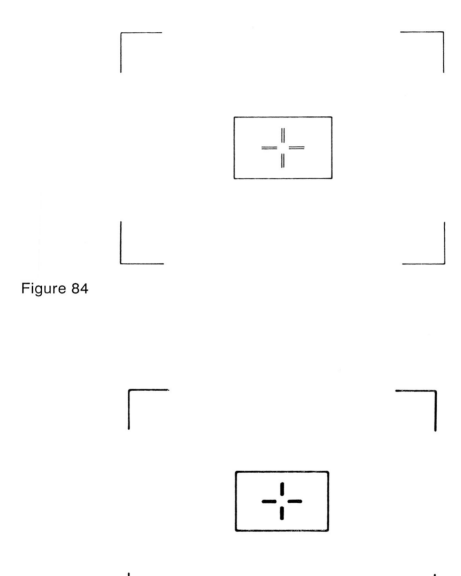

Figure 84

Figure 85

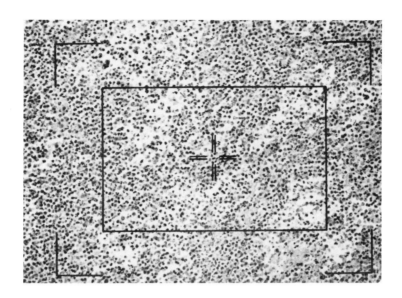

Figure 86

The adjustment of the focusing telescope should bring the reticle into sharp focus, as seen in Figure 84. Notice the double lines in the center cross. They should never appear as in Figure 85. The easiest way to accomplish this focus is to observe an evenly illuminated blank field since there are no interfering structures to hamper focusing the cross. Now place the specimen slide on the stage and bring it into focus using the coarse and fine adjustments on the microscope. Do not alter the reticle setting arrived at in the previous step. Use the fine adjustment to make certain the specimen is critically sharp. It is imperative that both the specimen and reticle appear sharp simultaneously (Figure 86). When using very low power objectives a double check on your focus may be made by using the parallax method. This is done by bringing the image into focus and then moving the head laterally back and forth to be certain the relative position of the specimen and cross lines remains fixed. In other words the specimen should not appear to move in relation to the reticle or *vice versa*. The foregoing procedure is the first step in making successful photomicrographs and is essential whether the film being used is black and white or color.

COLOR PHOTOMICROGRAPHY

Before we deal with color balance or exposure, it is necessary that we have some understanding of color films and what results we can expect to obtain from each type.

Different color films record certain biological stains with varying degrees of accuracy even when balanced for the same Kelvin temperature. Often there are broad spectral gaps where, green may appear blue or a vivid red brownish in hue. If a group of daylight balanced color films is selected and you use the recommended filtration for a specific light source, you will find that no two will record a given stained specimen the same way.

What are the most common causes of inconsistency in color reproduction? Exposure duration is one variable that must not be overlooked as a cause of variations and shifts in color. Some films are more sensitive than others to very long or short exposures. This is known as a films reciprocity failure and is a prime consideration in photomicrography. However, modern microscope illuminators and improved color films have greatly reduced this problem.

Another factor governing color results is the optical system itself. Even if the light source is perfectly balanced for the film and the exposure time well within the reciprocity tolerances, the color results may not be satisfactory. Some objectives tend to produce bluish tones while others may shift toward the yellow. This is not so much a factor with the newer optics but is a definite consideration with the slightly older ones. Should such a shift in color exist, it is easily remedied with a Kodak color correction (CC) filter. By making a series of test photographs using all the objectives in your battery, it will be a simple matter to determine which objectives, if any, have a tendency to shift in color. If the shift is toward the yellow, a CCB or blue filter should be used, if toward the blue a CCY or yellow filter should be chosen. Later in this chapter, the CC filters and a method of correction will be described.

As previously stated, many factors enter into incorrect color balance. One very common one is an improperly focused simple substage condenser. This can produce color shifts ranging from red through blue or violet. With objectives 10X and up the simplest solution is to use a aplanatic-achromatic condenser. However this condenser is not suitable for the low power lenses as it will not provide a full field of illumination.

The high degree of correction in this condenser completely eliminates the problem. Using the aplanatic condenser does not, however, eliminate the necessity for maintaining perfect focus as prescribed in the procedure for Köhler illumination. It does mean that a slight defocus will not affect the color balance of the photomicrograph.

Some of the color faults described are illustrated in Color Plate III.

All plates appear following page 135.

Remember when using a simple condenser always have the color fringes around the field diaphragm equally divided between red and blue to eliminate any serious shift in color. Also be certain to open up the diaphragm as described earlier, before making the photomicrograph.

Now let us turn to the most elementary part of the microscope, the light source. A wide spread of Kelvin temperatures is found among different lamps used for photomicrography. For example, Kodachrome 25, 64 or 200 are balanced for daylight (mean noon, 5,500° Kelvin). The manufacturer recommends a Wratten 80A or equivalent blue filter be used to bring 3,200°K to daylight temperature. Any lamp that does not burn at 3,200°K must be brought to that temperature by the use of an additional filter before adding the Wratten 80A conversion filter.

The older microscope illuminators used tungsten lamps and any filter recommendations given are for new lamps operated at their rated voltages. As a lamp is used, it gradually acquires a smoky film that reduces the color temperature to a degree proportional to the film's density. That is, the color temperature of the lamp drops continuously or gets yellower and will require additional blue filtration if carried to excess. Therefore, lamps should be removed and inspected periodically for such discoloration and replaced if necessary.

Newer microscopes are supplied with quartz-tungsten-halogen lamps. These lamps have many advantages. The most important of these is high intensity, small size, and consistent color temperature. Those designed for microscopy burn at 3,200°K. However, they do emit a blue band that interferes with color balance. This is easily remedied by placing a Wratten 2B filter in the light path. This filter should be left in the light path at all times and will efficiently eliminate the blue cast due to this anomaly. Unlike conventional tungsten lamps, this lamp does not change color temperature with use, and, any color balance established for a particular film is valid for the life of the lamp.

TABLE II. DAYLIGHT BALANCED FILMS		
Light Source	**Color Temperature degrees Kelvin**	**Filtration**
15W-T	2800°K	82C + 80A
20W-TH	2900°K	82C + 80A + CC1OB
60W-T	3050°K	82A+ 80A
60W-TH	3200°K	80A
100W-T	3150°K	CC20B + 80A
100W-TH	3200°K	80A

TABLE III. TUNGSTEN FILMS BALANCED FOR 3200°K		
Light Source	**Color Temperature degrees Kelvin**	**Filtration**
15W-T	2800°K	82C
20W-TH	2900°K	82C + CC20B
60W-T	3050°K	82A
60W-TH	3200°K	NONE
100W-T	3150°K	CC10B
100W-TH	3200°K	NONE
Tungsten Type A - Kodachrome Balanced For 3400°K		
15W-T	2800°K	82C + CC40B
20W-TH	2900°K	82C + CC20B
60W-T	3050°K	82B + CC10B
60W-TH	3200°K	82B
100W-T	3150°K	82A + CC10B
100W-TH	3200°K	82A

In all cases a 2B filter should be included in the filter pack when tungsten halogen (TH) lamps are used. 12V tungsten halogen lamps should be burned at 9V to yield 3200°K. Wratten 80A filters may be replaced with the Nikon NCB-10 filter. T = regular tungsten lamp. TH = tungsten halogen lamp.

Tables II and III list light sources, their color temperatures, and the necessary filtration to bring each to the proper color temperature for the type of color film being used. The filtration listed is intended as a starting point only and further CC filtration may be necessary to fine tune color reproduction.

Kodak Wratten filters are supplied only in gelatin and if used should be mounted in metal filter frames to avoid damage. Wratten equivalent filters mounted in glass may be purchased in any photographic shop under the TIFFENN name.

To intensify reds and blues in lightly stained specimens a 1 or 2mm Didymium filter should be included in the filter pack. However, this filter will oversaturate the colors if the stain is heavy; it should therefore be used with discretion.

CHOICE OF COLOR FILM

Ektachrome, Agfachrome or Fujichrome are good choices if quality in-house processing is available. In some areas custom professional processing laboratories are available for quality processing. Unless you wish to lose your mind with color variations, avoid amateur processing laboratories. Kodachrome may be processed by Kodak and by some professional laboratories. Kodak also processes Ektachrome, but if time is a factor, in house processing or a local custom laboratory is the only alternative. In the author's opinion the best film choices are Ektachrome T-50, Kodachrome A, and Kodachrome 64 or 25.

T-50 Ektachrome, unlike the others, is balanced for 3200°K. Therefore, it requires a minimum of filtration and is suitable for most photomicrography. While it does not have quite as fine a grain structure as Kodachrome, the difference is slight, and it is especially effective in the recording of biological stains. When higher emulsion speed is necessary there are a wide choice of emulsions available; an example is Ektachrome 1000.

TABLE IV. FAULTS IN COLOR BALANCE		
Appearance of Photomicrograph	**Possible Cause**	**Remedy**
Slightly yellow	Mounting medium Discolored lamp Incorrect voltage Processing fault	CCB correction Change lamp Adjust voltage Correct processing
Magenta tint	Processing fault Emulsion variance	Correct processing *See note below*
Blue tint	Abbe condenser improperly focused Over filtration Halogen lamp burned at too high a voltage	Adjust condenser Recheck filter pack Adjust voltage
Green tint	Heat absorbing filter in light path Residual stain in clear areas of slide Processing	CCM filtration CCM filtration Correct processing
Red tint	Abbe condenser improperly focused	Adjust condenser
Yellow-red	Daylight film with tungsten source—no correction	80A filter + 82 series depending upon light source used

* Film should be ordered in at least 20 pack batches, to insure having the same emulsion number, and refrigerated until at least two hours before use. Ask your supplier to be certain all film ordered has the same emulsion number. When a new supply is ordered make a test to check color balance.

USE OF CC FILTERS FOR COLOR BALANCE

CC, or color correction, filters are just what the name implies, *i.e.*, filters for fine tuning reproduction of colors with specific types of color films. Keeping in mind the preceding recommendations concerning color balance, color slides should be viewed and a determination made as to color tint in clear areas or any other inconsistencies readily apparent. Remember, the clear areas will not always be without tone as this will vary with the density of the stain in the specimen. However, if any tone does appear, it should be a neutral grey and not reveal any color tints.

After making your initial test using the recommended filtration, the slides should be checked for color accuracy. To do this, the color print viewing kit manufactured by Eastman Kodak should be used. Figure 87 shows the kit of filter

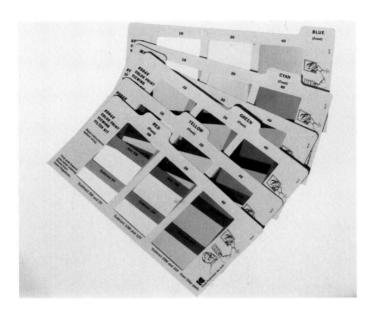

Figure 87

cards ranging from the additive primary colors, red, blue and green to the subtractive compliments magenta, cyan and yellow in units of 10, 20, and 40. Complete information and instructions are included with each set.

The first step is to select a color film. If you select T-50 Ektachrome, for instance, check with the manufacturer and find out at what color temperature the lamp in your microscope operates using the recommended voltage. Chances are that it will approach 3200°K. The table in this chapter should include the lamp you are using and will negate the necessity of checking with the manufacturer.

With the newer, halogen lamp equipped, microscopes this is not a problem since most burn at 3200°K. Using the information in the foregoing table for T-50 film, and varying the ASA from 50 to 32, expose a test roll. Choose a specimen with some clear areas to permit easier checking of color tints. I recommend that you have the film processed by Kodak or a local custom professional laboratory to obtain a standard and keep careful exposure and filtration records. When you receive the processed film, place each transparency in numerical order so that exposure and filter data may be checked against notes taken when the exposures were made.

Figure 88

Using a light box with a tungsten illuminating system that is, or approaches, 3200°K, examine your slides for color balance. This is essential since the slides will be projected at that temperature. Unless a light box approaching this tempera- ture is used it will be impossible to balance your color. Never use an X-ray viewer or fluorescent light to balance color.

If the best exposure happens to have a blue tint, use the filter kit to see what correction is necessary to obtain a neutral background (Figure 88). In this case you would choose the compliment yellow card. If the transparency looks normal with the CC20Y, expose another test roll using the CC10Y for half the exposures and the CC20Y for the remainder. In making this test, vary the ASA rating from 32 to 50. In Figure 88, the filter card is shown being held in front of the transparencies. This is for making a preliminary check. The final determination is made holding the transparency in front of the filters.

To use the filters for correction, when photographing place them over the light port of the microscope, as shown in Figure 89. Keep careful notes and send the film to your chosen laboratory for processing. It must be pointed out here that, for consistent results, you should always use the same laboratory. When the film is returned, one ASA rating and filter combination should result in a perfect balance. This is the combination you should stay with when making all your color photographs with Ektachrome T-50 film.

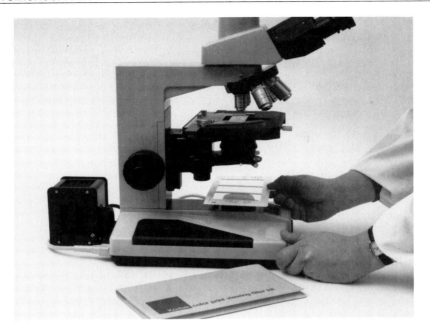

Figure 89

The same procedure would apply if you were to choose another film such as Kodachrome Type A. This is another fine color film for photomicrography but, because it is balanced for 3400°K, an 82A filter must be used with the 3200°K light source. After making an initial test, the same procedure should be followed as described for Ektachrome.

The following will help to understand what the CC filters do for color correction.

- Cyan removes red
- Blue removes yellow
- Green removes magenta
- Red removes cyan
- Yellow removes blue
- Magenta removes green

If the above preceding procedures are followed, there should be little difficulty in fine-tuning your color transparencies.

Before we deal with black and white photography, it is necessary to understand how to control light intensity, regardless of whether color or black and white film is being used.

Some modern photomicrographic cameras use a go-no-go system. This deactivates the shutter if there is too little or too much light for the shutter speed range. There is an over-ride mechanism for manual exposures if this becomes necessary. For the most part, too much light is the problem in brightfield illumination. The light intensity must not be controlled by varying the lamp voltage. Brightness is controlled only by using neutral density filters in the light path. Kodak supplies these in gelatin for light attenuation throughout the entire visible spectrum. They are available in 12 densities, with a transmittance range of 80 percent to 0.01 percent. These filters though they are called neutral, do not react that way with color films. They produce a warm or slightly yellow cast to the transparencies. Therefore, the best choice is the filters offered by the microscope manufacturers. These glass filters are made by the evaporation process and are truly neutral. All color balancing filter recommendations in the forgoing tables were determined using these evaporation coated glass filters.

BLACK AND WHITE PHOTOMICROGRAPHY

The first consideration when making photomicrographs is a choice of film. In black and white there are many films to choose from. For most applications the slower, high resolution films such as T-MAX 100, Plus-X, Technical Pan, and Ilford Pan-F will satisfy most needs. When high speed is required for very low level light conditions there is Kodak TRI-X, T-MAX 400, and T-MAX P-3200. When several camera backs with dark slides are available, each may be loaded with a different emulsion. This will enable the user to choose the film/developer combination most suitable for the specimens. If one is limited to a single camera back, a compromise must be reached by choosing the film type and developer best suited for the majority of the slides to be photographed. The same compromise must be made when choosing a developer. Developers range from high energy, high contrast developers such as D-19 to the so called "softer working" developers DK-50, D-76, and HC-110. The role the film, developer, and filters play in the final results, cannot be over emphasized. Films vary in their inherent contrast characteristics and can be further manipulated by the developer chosen. Contrast should always be controlled photographically, *i.e.*, by film and developer. Color differentiation is controlled with filters. In some instances, it will prove most satisfactory if specimens are grouped according to stain density, *e.g.*, light, normal, and dense, and photographed accordingly. This procedure is recommended when there is a wide variety of stain characteristics. Naturally, if the stains are more or less uniform a single procedure may be followed.

What do the filters do and how should they be chosen? This is a complex subject but the intent of this book is to simplify. The proper choice of filtration is determined by the color of the stain. What should be avoided, except in special cases, is overfiltration. Remember, the purpose of making a photomicrograph is to show all the details in the specimen. For example, when choosing a filter for a specimen with a red stain, be certain that the red areas do not appear black. The green filter chosen should be of average density such as a Wratten No. 11 or 13. Both of these filters will transmit red so that visually it is possible to recognize red areas as red not black. Table V lists filter choices for different colors.

TABLE V.	
Stain Color	**Filter No. And Color**
Red	Wratten No. 11, 13, 58 (light to dark green)
Green	Wratten No. 25, 23, (red to orange)
Blue	Wratten No. 8, 15, 23 (light yellow to orange)
Yellow	Wratten No. 47 (blue)
Orange	Wratten No. 47 (blue)

A complete set of filters for photomicrography is available from Eastman Kodak and can be used singly or in combination to provide complete control of differentiation in photomicrographs regardless of the density or type of stain used. The suggested selections are as defined in Table VI.

TABLE VI.		
Filter No.	**Filter No.**	**Filter No.**
11	25	47
12	29	47A
13	35	47B
15	45	58
22	47	61

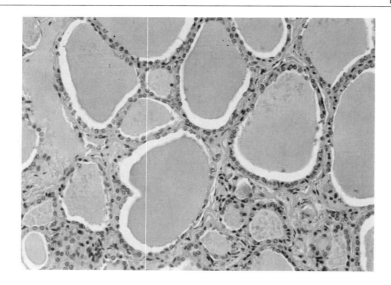

Figure 90

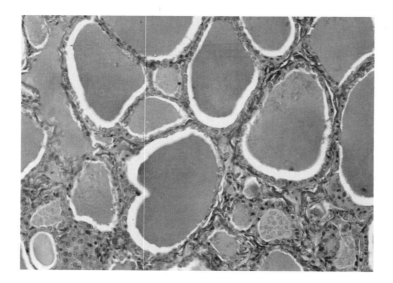

Figure 91

We have seen the results of correct and of overfiltration in Figures 90 and 91. Now let's look at another situation. How filters can work for you or against you is illustrated in Figure 92. The purpose of this photograph is to demonstrate the presence of bacteria in the tissue. The liver section was stained red and the bacteria were very deep purple, almost black. The normal procedure for a pink or red stain

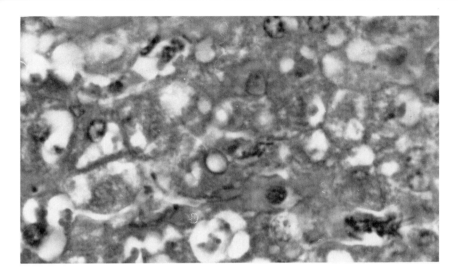

Figure 92

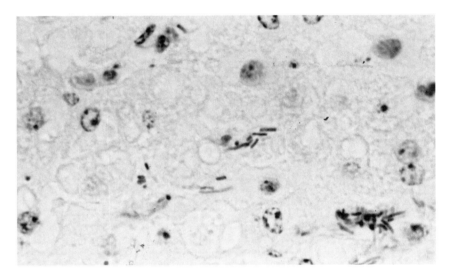

Figure 93

would be to use a green filter. Here the green filter was used and the primary focus was on the bacteria. The result is a useless photomicrograph since the green filter darkened the tissue to such an extent that the bacteria are all but invisible. The same section is shown in Figure 93. Here a red filter was used to minimize the tissue and accent the bacteria. The red filter passes the red making it dark on the negative and light on the print.

Filters can either subtract or add density on the final print. For example, if we have a red specimen and we would like the final print to have good contrast and show cellular details, a medium green filter should be used. This will make the red tissue a medium grey in the final print. Why? Because the green filter will pass only a portion of the red and block out the rest, resulting in a medium grey tone in the final print. However, if there are structures in the red areas which will normally reproduce as a grey tone, we will want to minimize the red areas in order to bring out the desired structures. To do this, we should select either a deep orange or red filter depending upon the intensity of the red stain. By doing this, we will allow the red to pass on to the negative, producing a dark image. This in turn will finalize as a light image in the positive print. Color plate IV in the color section graphically illustrates this point. Filters must be chosen with great care. Unfortunately, the green filter supplied with the microscope is often the only one ever used and this results in many unsatisfactory photomicrographs.

FILTER FACTORS

While the cameras supplied today for photomicrography are claimed to be automatic, there is a big loophole in this claim. There are two types of light sensors used to detect the amount of light reaching the film plane to control the length of exposure; photomultipliers and silicon photodiodes. Both have different sensitivity and spectral response characteristics. The photomultiplier is more sensitive to low light level situations such as encountered in spot metering and fluorescence photography but its spectral response leaves a great deal to be desired. All the filters listed have factors which must be applied when making an exposure. In black and white photomicrography using narrow band filters of various wavelengths, the automatic exposure camera is no longer automatic. In other words compensation must be made for certain filters. For example, if you are using black and white film with an ASA of 100 and set the ASA on the control box of the camera at 100 and make a photograph using a No. 11 or 13 green filter the exposure index will be valid and a perfect exposure will result automatically. However, if a No. 25, 47, or 33 filter is required the ASA setting of 100 will no longer be correct. Manufacturers do not supply correction factors for the narrow band filters nor are they consistent from one manufacturer to another.

A No. 25 filter has a factor of 8 with T-MAX 100 film. In conventional photography this means that the camera lens aperture would have to be opened up 3 f:stops to compensate for the filter. A one second exposure would become an 8 second exposure if the lens were not opened up 3 f:stops. With the photo micrographic camera we are not dealing with f:stops. We must therefore compensate in another way. Instead of a setting of 100 we would have to use an ASA setting of 12 to obtain a normal exposure. Unfortunately, with some cameras the sensitivity adjustment has been made in the opposite direction and an ASA of 100

All plates appear following page 135.

would require a setting of 800 in order to obtain a correct exposure. In the author's opinion the latter is a more desirable direction for the manufacturers to follow. It is indeed unfortunate that there is no standardization among manufacturers. The only solution for the user is to make a series of tests and record the results, so that a set of factors may be applied to your particular camera.

With the silicon diode sensor this problem is much less bothersome, as its spectral response is flatter than the photomultiplier's. Unfortunately, the silicon diodes do not have the high sensitivity to low levels of illumination, nor do cameras using this system have spot metering. With many users these factors will be of no consequence and this would be the way to go. Regardless of the system chosen, careful tests are necessary.

SELECTING THE PROPER BLACK AND WHITE FILM

Selection of the proper film is very important and to a large extent depends upon the nature of the material being recorded. Early in this chapter it was stated that we would deal primarily with the 35mm format because it was the hands-down choice among those engaged in recording images generated by the microscope. Its use for color requires no explanation but, for black and white nothing equals the large 4 X 5 format. It is for those few who have large format capability that some 4 X 5 films and developer recommendations will be listed. The following is a list of films the author considers to be the most suitable for a major percentage of applications.

35 mm

Kodak Panatomic X, ASA 50 [1]

This is a very fine grain film with excellent sharpness and resolution and is well suited to most brightfield work. Suggested development in *fresh* KODAK DK-50 for 5 min. at 70°F with intermittent agitation.

Kodak T-Max 100

A new generation of KODAK films possessing very fine grain, sharpness and resolution. It is suitable for all brightfield work and its speed of 100 also makes it a fine choice for phase contrast and DIC photography. There are a number of development choices for this film, for example per the following.

DK-50 5 min. at 70°F intermittent agitation.

D-76 9 min. at 70°F intermittent agitation.

T-Max 6½ min at 75°F intermittent agitation.

HC-110 Dil.B 7 min. at 70°F intermittent agitation

---

1   Although this film is listed as having an ASA of 32, its practical working ASA with the above recommended development is 50.

The author's personal preference is DK-50. It produces a contrast range that will keep paper contrast between a 2 and 3 printing filter or contrast grade.

Kodak Plus-X

This film has fine spectral response and a speed of ASA 125. The grain structure is fine and resolution and sharpness are excellent. Its speed makes it suitable for bright field, phase contrast and DIC. While the same developer choice applies as for T-MAX, the authors choice is DK-50 for 5 min. at 70°F. DK-50 is a clean-working, no fuss, long lasting developer that seldom fails to produce a high quality negative.

Kodak T-Max 400

A fast (ASA 400) film with fine grain and excellent sharpness. Well suited for fluorescence or low light situations. May be push processed to ASA 3200. Push processing should only be used in extreme conditions, as the push process results in excessive grain. Normal processing time in T-MAX developer is 6 min. at 75°F. with intermittent agitation.

Kodak Technical Pan

This film has exceptionally fine grain, very high resolution and sharpness. Its speed range is from ASA 25 to 125 depending on the developer chosen. It also has a wide contrast range which also depends on the developer chosen. Kodak has formulated a special developer for this film called TECHIDOL. It comes in both powder and liquid forms and the developing procedure for each is entirely different. With TECHNIDOL-LC (powder form) agitation is carried out by gently rotating the tank so that the tank is inverted with each rotation. If on the other hand the liquid concentrate is used the agitation is straight up and down (do not rotate the tank). Shake the tank up and down 10 to 12 times in no more than 2 seconds. Let the tank sit for 28 seconds and repeat the next 2 second agitation. Repeat every 30 seconds for the remainder of the development time. For phase contrast and DIC work and an ASA of 125 the author uses DK-50 for 5 minutes at 70°F. This combination is also good for brightfield work. D-76 and HC-110 are also excellent developers for the film and which is the more suitable will depend upon the individual application

PROCESSING

The processing of film is what most people do not like to do and their results show it. No matter how critical you are about microscope alignment, filtration, exposure, and choice of film, it will be for naught if the processing is slighted. Processing is a critical operation and there are no short cuts. For the most part, films are processed in modest facilities using stainless steel reels and cans. Hence, that is the method which we will discuss here. This method produces excellent results but must be carried out with extreme care. First it is important that all solutions are at the same temperature. If 70° is chosen as the developing tempera-ture, the developer, rinse, hypo , and wash should be held at this temperature. After winding the film on the reel, place it in a can of suitable capacity. In other words, if one roll is to be processed, use a small two reel can. Be certain under these circumstances to place one dummy reel in the can along with the film reel. This prevents the loaded reel from excessive up and down movement when the can is inverted, thereby preventing the developer from jetting through the perforations. The jetting action causes dark streaks to emanate from the perforations, making acceptable prints impossible. The following is an agitation procedure that will apply to all films except TECHNICAL PAN when processed in liquid TECH-NIDOL. Agitate the film every thirty seconds by picking up the can, rotate slowly left to right, invert and twist left to right and gently place back on work bench. This method produces perfect and consistent results and eliminates a major variable in photomicrography. When development has been completed, pour out the developer and fill the can with water of the same temperature. Agitate the water in the same manner described for development, except here the action should be constant. Rinse for one minute, decant the water, and pour in the hypo. Agitate constantly for the first minute and complete by intermittent agitation for ten minutes. The film should then be washed for thirty minutes at the same temperature, sponged and dried at a moderate temperature. All the forgoing is based upon the use of fresh chemistry. Exhausted solutions are the cause of most processing failures. As a rule, one should follow either the one shot route or use 25-36 exposure rolls per gallon. Twenty-five rolls per gallon is borderline but will provide high quality negatives provided the developer is not stored more then three weeks. Developer should not be stored with an air space at the top of the bottle. Storage should be in accordion bellows bottles so that all air may be expelled before capping. It may seem elementary but oxidized developer is the cause of more failures in light and electron micrography than any other single factor, except, of course, not knowing how to use the microscope.

LARGE FORMAT FILMS (4X5)

Kodak Ektapan ASA 100

Develop in DK-50, 5 min. 70°F. Agitate, 5 seconds every 30 seconds by lifting the film hangers from the developer and draining from alternate corners. For higher contrast, develop in Kodak D-11 for 5 min. at 70°F and use an ASA of 200.

Kodak Technical Pan

Process in TECHNIDOL-LC developer for 5 min. at 70°F. and use an ASA of 25. This will produce a long scale negative and is ideal for contrasty specimens. For a higher contrast use DK-50 for 5 min. at 70°F. Agitate as directed for EKTAPAN. Using DK-50 will boost the ASA rating to125.

Technidol-LC is the powder form of this developer and is the only type recommended by the author.

CHAPTER 9

FLUORESCENCE MICROSCOPY

To this point we have covered the most widely used contrast methods employed in visible light, but we will now discuss a method that takes advantage of invisible or near-invisible regions of the electromagnetic spectrum.

Because of the many areas where fluorescence microscopy is used, and varying individual needs, it is impossible, and not in keeping with the purpose of this book, to discuss tissue preparation technology for fluorescence microscopy.

In fluorescence microscopy short wavelengths of the spectrum, ranging from the invisible to the green portion of the spectrum, are used to excite dyes (fluorochromes) or naturally fluorescent specimens into self-luminous objects. This self-luminance is the result of the release of energy in the form of visible light. Herein lies the difference between fluorescing preparations and the general method of microscopy in which transmitted light is observed and photographed.

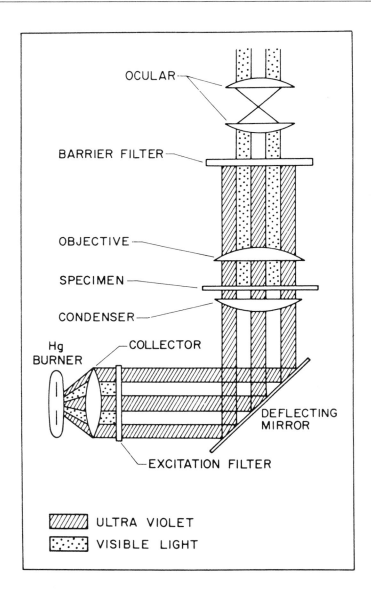

OCULAR

BARRIER FILTER

OBJECTIVE

SPECIMEN

CONDENSER

Hg
BURNER

COLLECTOR

DEFLECTING
MIRROR

EXCITATION FILTER

ULTRA VIOLET
VISIBLE LIGHT

Figure 94

In fluorescence, specimens become self-luminous and, as explained earlier, this is the result of the absorption of short wave radiation. The short-wave radiation exciting the fluorochrome into luminosity does not contribute to the formation of the image. These wavelengths are absorbed by the barrier filter. Figure 94 graphically demonstrates the principle of fluorescence microscopy. Notice that the products of light waves are all in the visible range. It is only the light waves resulting from excitation that form the final image.

Figure 95, shows a modern fluorescence microscope. The design has been vastly improved over the past decade and simplicity is the key word for this present state-of-the art instrument. The instrument shown has capabilities in both transmitted and reflected light. Both methods have areas in which they excel and, therefore, if a wide scope of work is to be undertaken, the acquisition of such an instrument is advisable.

PERTINENT APPLICATIONS

Due to their ease of detection fluorescent substances are of particular interest in the study of the distribution and absorption of drugs in tissue.

When infectious agents called antigens are introduced into a living organism, substances are produced which react with the alien materials. These substances are called antibodies and can be isolated and purified and reacte with its corresponding antigen outside the body as well. For instance, a fluorochrome dye can be coupled to an antibody and detected in the fluorescein-isothiocyanate (FITC) fluorescent dye system.

What are some of the optical aspects that should be considered? Some thought should be given to the type of specimen slide and cover slip used. For routine work in transmitted light an ordinary specimen slide and cover slip will be found to be fairly satisfactory. In epi or reflected light the specimen may or may not be covered, so that slide and cover slip are not a consideration. However, most clinical fluorescence kits utilize cover slip covered specimens, and the slide and cover slip will influence results. For transmitted light where critical research work is being done Corning Corex D glass is an excellent choice. Conventional slides will fluoresce, resulting in a loss of ultraviolet light and a decrease in contrast between the self-luminous object and the background. Compatible cover slips are also available but ordinary cover slips are satisfactory for most work.

EPI-FLUORESCENCE

The epi or reflected light system is presently favored for most studies. Hence, this is the system which receives emphasis here. Reflected light fluorescence has the advantage of high efficiency and simplicity of operation because the objective serves as its own condenser. An example is the Nikon Epi-Fluorescence attachment "EF" shown in Figure 95. This instrument features a 50W AC high pressure mercury illuminator with exciter-barrier dichroic mirror combination for UV B and G ranges. If one is working with FITC exclusively, the Nikon EFA would be the logical choice. This consists of a 12 volt 50 watt halogen illuminator with an exciter barrier=dichroic mirror combination for the B excitation range. These filters and light sources are available from most manufacturers and the choice would be dictated by the make of instrument being used. A special series of UV F

Figure 95

glycerine objectives has been computed by NIKON that permits higher transmission and numerical aperture, thereby shortening exposures and minimizing quenching or fading of the fluorescent image.

An anti-quenching formula that has been passed on to the author is as follows.

MATERIALS

DABCO 1,4 Diazabicyclo (2,2,2) Octane (Triethylene diamine). Eastman Organic Chemicals. Cat. # P8076. May be ordered from American Scientific Products, Catalog # G 279.

GELVATOL—20-30 (Polyvinal Alcohol) Monsanto

PREPARATION

(1) Suspend 20G of Gelvatol in 80ml of phosphate buffered saline (0.15 M). Adjust to pH 7.2. Stir for 24 hours on magnetic stirrer at room temperature.

(2) Add 20ml of glycerol and continue stirring for 24 hours.

(3) Remove undissolved Gelvatol 20-30 by centrifugation at about 12,000 rpm for 30 minutes.

WORKING MEDIUM

To 100ml of the above add 3.37 gms of DABCO. Be sure the Gelvatol is allowed to reach about 70°F before adding DABCO. This combination must be stabilized at a pH of 8.6

STORAGE

Store in airtight bottle at 4°C. Record pH and label bottle. pH should be 8.6. This mounting medium has excellent keeping qualities.

PHOTOGRAPHY

As the final product of a study is, for the most part, a photographic image, the following details will help suggest the most efficient methods available.

THE CAMERA

There are many different types of cameras available for photomicrography but, for fluorescence photomicrography, the ideal features are, spot metering, memory, and double exposure capability. With spot metering the sensor looks only at the self-luminous area and not at the black background. This makes possible very accurate exposures, even at low light levels. The double exposure feature allows two fluorochromes excited at different wavelengths to be recorded on a single frame of film. It further permits two illuminating techniques to be recorded as a composite. A typical example would be phase contrast in transmitted light and fluorescence in eip-illumination. To ensure the shortest possible exposure times there should be a minimum of glass between the projection ocular and the film.

TABLE VII. BLACK AND WHITE FILMS		
KODAK TRI-X	ASA-400*	DK-50, 5 min. 70°F, or D-76, 8 min 70°F, intermittent agitation
KODAK TRI-X	ASA-800	DK-50, 8min. 70°F.
KODAK-TRI-X	ASA-1200	D-19, 9 min. 70°F.
T-MAX-400	ASA-400/800	T-MAX, 7 min. 70°F.
T-MAX-400	ASA-1600	T-MAX, 9 min. 70°F.
T-MAX-400	ASA-3200**	T-MAX, 9.5 min. 75°F.
* ASA 400 is the best choice for a quality negative.		
** Use only under the most extreme cases. Grain and quality are poor.		

FILM

Color film is the obvious choice and in some cases it is the only choice. However, the cost of color plates for publication is not always compatible with many budgets. Therefore, for situations where black and white photographs will suffice, the most suitable films and developers are listed in Table VII.

There is another black and white film not included in the forgoing list; it is called Illford XP-2. It has what is called a chromogenic emulsion because of its multilayer construction and because it is processed in C-41 color developer. It has a grain structure comparable to Panatomic-X or Pan-F and an incredible ASA rating of 400. It has an extremely broad tonal range and will capture the most subtle tones that fluorescence can produce. It may be processed by the user or sent to any independent color laboratory for C-41 processing. The C-41 process is the designation for processing Kodacolor or Vericolor films. Ilford supplies a kit for processing this film that consists of two solutions, a developer and a bleach-fix or "blix". This developer is very simple to use but, after processing and washing is completed, do not attempt to sponge off the film prior to drying. The emulsion is quite delicate when wet, therefore, immersion in a wetting agent such as Kodak Fotoflo is recommended. Hang up the film and let dry in dust-free warm (not hot) air.

COLOR FILMS

Ektachrome 200

This is the author's choice for fluorescence photomicrography. Its ASA of 200 is suitable for most applications but if necessary it may be push processed to 400 with excellent results. It produces bright colors and has very fine grain, high resolving power and sharpness.

Ektachrome P800/1600

Where extreme conditions are encountered this film performs well. It may be processed at an exposure index of ASA 800 or 1600. Grain is marginal fine, sharpness is good and resolving power is medium.

It must be remembered that in black and white or color photography, you never get something for nothing. What one loses in the push process is grain structure, but, when speed is essential, naturally,there is only one route to take. The new cameras described will help solve this problem since less glass in the system means finer image quality and a greater amount of light reaching the film.

PRECAUTIONS

Never view a field radiated by ultraviolet light without the barrier filter in place. Such observation may result in permanent eye damage. I repeat, never.

CHAPTER 10

FIELD LOCATION

When scanning a slide the microscopist often finds that areas of interest are located in many parts of the specimen. If the same microscope is used at all times, relocation does not present a problem since all one has to do is read the X and Y scales on the stage for pinpointing each area. However, this is not as simple if different scopes are used, particularly if they have been produced by different manufacturers. Conversion factors may be used but this is a nuisance and is time consuming.

Figure 96

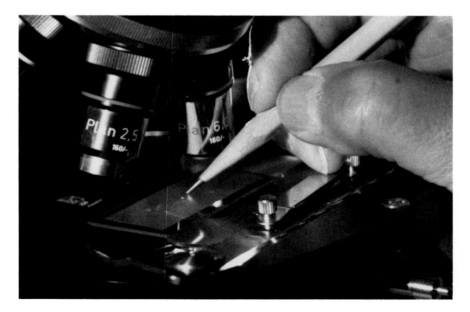

Figure 97

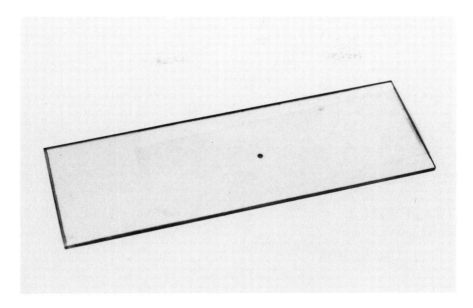

Figure 98

A simple method will be described so that this often-confronted problem may be avoided. First, be certain that the microscope is perfectly centered and that Köhler illumination has been achieved. When this step has been completed, place a strip of Scotch Brand Magic tape on a blank slide (Figure 96). Be sure to label the blank as you have labeled the one to be examined. Next, when the area of interest is located, replace the specimen slide with the blank. With the blank slide in place on the stage, stop down the radiant and aperture diaphragms fully. This will project a pinpoint of light on the tape. Using a very sharp lead pencil, or, preferably, a needle, mark or prick the spot depending on the choice of markers (Figure 97). This will leave a spot on the tape as seen in Figure 98, exaggerated here for the sake of illustration. Now the marker slide may be replaced with the specimen slide and the procedure repeated if another area of interest is located. The spots may be numbered 1, 2, 3, etc. for recording interesting features so that a number can be related to a specific area in the section or smear. It is wise to use a low power for this procedure. My choice is a 4, 6.3, or 10X objective. The low power makes it very simple to locate the mark on the location slide. Use the same low power to pick up your relocated area, center it, and increase magnification as dictated by the specimen.

TABLE VIII.	
Color as Seen in White Light	**Colors of Light Absorbed**
Red	Blue and Green
Blue	Red and Green
Green	Red and Blue
Yellow (Red plus Green)	Blue
Magenta (red-blue)	Green
Cyan (blue-green)	Red
Black	Red, Green, and Blue
White	None
Gray	Equal Portions of Red, Green, and Blue

CHAPTER 11

SPECIAL TECHNIQUES

DIFFERENTIAL OPTICAL STAINING USING FIBER OPTICS AND THE ADDITIVE COLOR TECHNIQUE

To understand what makes this technique possible, it is necessary to understand the nature of light, color, and color vision.

When all of the wavelengths between 400 and 700mm are present in nearly equal quantities we perceive colorless or white light. There is no absolute standard for white without a reference. In other words, we adapt quickly to any reasonable uniform distribution of energy in the prevailing illumination. For example, when there is little or no daylight present for comparison, we accept tungsten illumination as white in spite of the fact that there is far less blue and more red than in daylight. In a room dominated by daylight, a tungsten lamp appears distinctly yellow because we are now adapted to daylight.

In order to understand how the human eye sees colors let us consider the action of light filters as explained in Chapter 8 and illustrated in Color Plate IV. When white light, which consists of RED, BLUE, AND GREEN falls upon a filter which we term RED, it absorbs (subtracts) BLUE and GREEN light, and therefore appears RED. Similary a piece of red paper appears RED because it reflects RED light to the eye and absorbs BLUE and GREEN.

To further simplify the understanding of the phenomenon of color perception, Table VIII. and the illustrations in Color Plate V should eliminate doubts or questions that would hinder the understanding of optical staining.

Color plate V demonstrates graphically the composition of white light. The flask was filled with a milky solution to act as an integrating body. The three additive primary colors red, green, and blue were added one at a time until all three were shining on the flask. The results are No. 1 = red; No. 2, red + green = yellow; and No. 3, red + green + blue = white. That red and green = yellow is surprising at first. However it is easier to understand if we bear in mind that this is not yellow as seen in the spectrum. Spectral yellow is composed of a narrow band of wavelengths between 575nm and 590nm. What we see here is a broad band of wavelengths which includes all wavelengths of light except those in the blue region of the spectrum. By using two light sources, one with a green filter which transmits no wavelengths of light longer than 575nm and the other with a red filter which transmits no wavelength shorter than 590nm, we obtain the sensation of yellow

All plates appear following page 135.

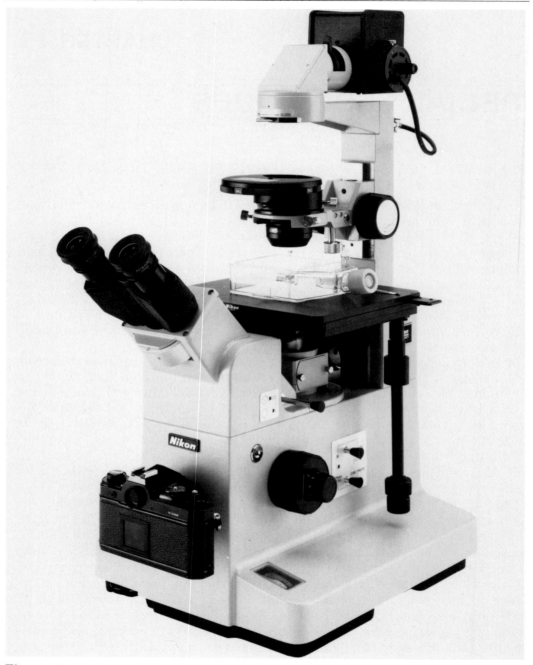

Figure 99

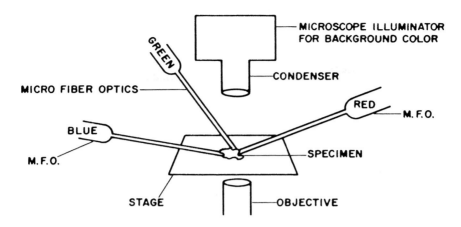

Figure 100

without using any of the wavelengths which appear yellow in the spectrum. The essential factor is the equal stimulation of the red and green receptors. By mixing red, green, and blue in varying proportions practically all colors can be produced and this is the basic principle of additive color optical staining.

The specimens we view and photograph using the microscope vary in their optical characteristics, namely absorption, surface characteristics, interference, and dispersion. In differential optical staining we take advantage of all the forgoing and regulate many by the angle of illumination.

Fiber optics have long been used with stereo microscopes in the biomedical and industrial fields. The fiber bundles, or light pipes as they are often called, provide a very brilliant, maneuverable, low heat light source. When applied to the conventional compound microscope, their application has been limited because of the large diameter of the light pipes and the extremely short working distance of conventional microscope objectives. An inverted microscope such as shown in (Figure 99) is the ideal instrument due to its very long working distance condenser which provides ample space for placing extra light sources at critical angles as shown in Figure 100. Red, blue, and green filters punched out of a filter sheet with a loose leaf punch are placed in the adapter sleeves of the 1mm fiber optic needles, The adapter sleeves fit over the standard 5mm diameter light pipes, thus reducing the diameter of the light by a factor of five. This size reduction enables the Red, Green, and Blue micro bundles to be brought very close to the specimen, and also makes possible immersion when specimens are in a liquid medium.

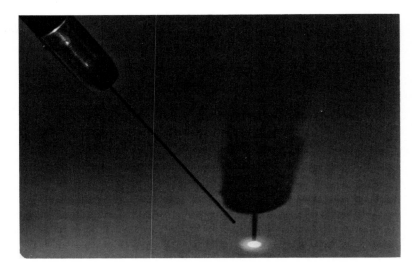

Figure 101

Figure 102

Figure 101 shows one type of micro fiber optic. Notice the intense, concentrated spot of light produced. The micro fibers may be used as a single source, filtered or unfiltered as the situation dictates, or in a triple red, green, blue combination for optical staining, as shown in Figure 100. The light exit of a triple set of fibers is shown in Figure 102. The micro fibers as described are furnished by American Volpi Corp.

Figure 103

For those having fiber optics of standard design with larger diameter, the application is somewhat limited for physical reasons. However, there are still many instances where they may be applied. A typical set of fibers, as supplied by Nikon, is shown in Figure 103. To Red, green, and blue filter these the author has found the following to work quite well. Cut off plastic soda straws to a length of one inch. These will slip over the end of the Nikon fibers. Next from red, green, and blue filter sheets, using a loose leaf paper punch, punch out a filter of each color. Now using Duco cement or airplane glue attach a filter to the end of each straw (Figure 104). These may now be slipped over the ends of the fibers in any combination desired for optical staining.

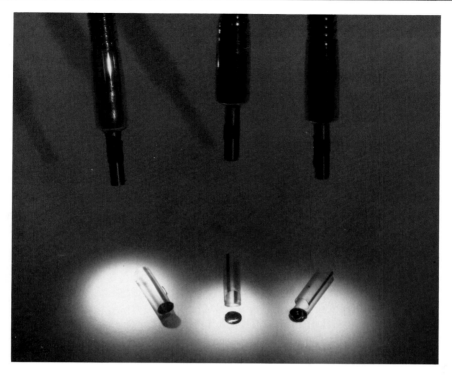

Figure 104

The background may be a complimenting color, black, or white as suits the specimen. The color intensity of the background is controlled by varying the voltage of the microscope illuminator. Color gels for this purpose may be cut to fit the filter holder of the microscope light source.

MICRO REPLICATION

Micro replication is a method for making thin, transparent plastic reproductions from the surfaces of objects which, because of their shape, size, weight or location, cannot be placed under a microscope. One of the more salient features of the process is its ability to provide a means of observing and photographing curved surfaces, such as cylinder walls or bearings, in a flat plane. This feature also makes possible serial photomicrographs which can be butted together so that a 360° surface may be laid out and studied in minute detail and charted as to wear, *etc.* Such examination does not require cutting or altering the specimen in any way. If found to be free of excessive wear it may be replaced in the machine for continued operation. The process is not limited solely to the study of metal surfaces. It is also useful in biology and medicine. The microscopist will soon become aware, after a little experience has been gained, that one application usually suggests another, with more and varied objects.

Materials And Methods

The materials for making micro replicas are not complicated or expensive. Figure 105 shows all the materials necessary for successful replication. Shown from left to right are: acetone for cleaning the surface to be replicated; sable brush for applying methylethylketone (MEK) solvent; a beaker containing MEK; scissor for cutting tape; a roll of replication tape (this is the same tape used in electron microscopy); a medicine dropper; scotch tape for fixing the replica on a microscope slide; and a glass plate to provide a hard smooth surface for rolling or pressing an object on replication tape. In Figure 106 acetone is being applied to the surface of a bearing to remove all traces of dirt or grease. The next step, (Figure 107) shows the replication tape being cut. In this case the tape will be cut slightly longer than the circumference of the bearing. This is necessary so that after applying MEK with the brush (Figure 108) the tape may be cinched tightly around the surface to insure good contact (Figure 109). The tape should remain in contact with the surface for at least one minute, or long enough to allow partial hardening as the solvent evaporates. In (Figure 110) the replica is mounted face down and firmly affixed to the surface of a 1" X 3" microscope slide by means of two strips of scotch tape at either end.

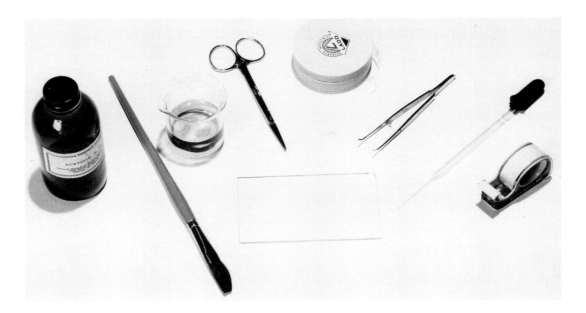

Figure 105

Figure 106

Figure 107

Figure 108

Figure 109

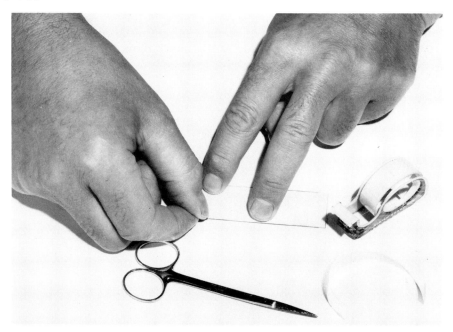

Figure 110

Microscopy

The illumination employed is of the utmost importance in order to bring out the fine microstructures which have been transferred to the plastic tape. To enhance the texture and produce a pseudo three-dimensional effect, oblique illumination is used, as described in Chapter 4.

Figure 111 shows a replicated bearing surface magnified 300X. Score marks and some evidence of chatter can be detected. After further operation (Figure 112), scoring is much deeper and chatter (vertical lines) is very apparent. Figure 113 shows the same bearing surface after further use. Notice the heavy scoring and deep gouging which resulted in a complete breakdown of the machine. The forgoing photomicrographs were made as previously stated using oblique illumination. There is, however, another type of transmitted illumination equally useful in the photomicrography of replications. Figure 114 shows the surface of a bearing magnified 500X. The texture in this case was brought out by the use of differential interference contrast (Nomarski). Figure 115, is a replication of the skin from the inner surface of the forearm magnified 400X using Nomarski DIC. These are but a few of the many applications which lend them selves to the process.

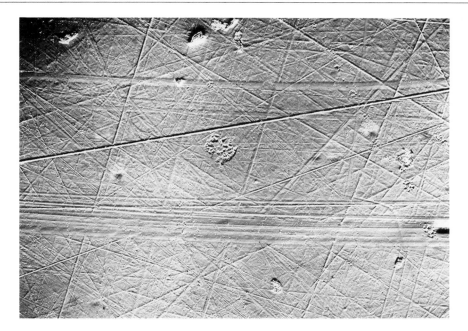

Figure 111

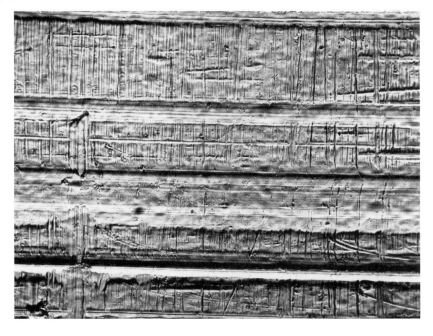

Figure 112

Figure 113

Figure 114

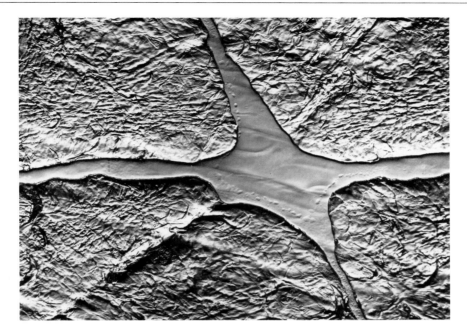

Figure 115

Photomicrography

Objectives used were of the plano or flat field type. Monochromatic green light produced by an interference filter with a peak transmission of 546nm was used in all cases regardless of the spherical and chromatic correction of the optical system. The film found best suited for the photomicrography was Kodak Contrast Process Ortho. When developed by inspection in fresh DK-50 the contrast range can be easily controlled within 1, 2, and 3 printing grades of paper.

To summarize, high resolution photomicrographs may be made from replicas of almost any clean, dry surface. Two methods of illumination, oblique and Nomarski interference, may be employed with about equal success.

SOURCE LIST

VOLPI FIBER OPTIC ILLUMINATOR: American Volpi Corp., 26
Aureilus Ave., Auburn N.Y. 13021-0400. 315-253-9707.
COLOR GELS: Edmund Scientific Co., 1010 Glouster Pike, Barrington
N.J. 08007. 609-547-3488.

REFERENCES

Differential Optical Staining Using Fiber Optics and the Additive Color Technique

Smith, Robert F. Differential Optical Staining of Colorless Living Organisms in Macro Photography. *Journal of the Biological Photographic Association.* May-Aug. 1955

Smith, Robert F. Color Contrast Methods in Microscopy and Photo micrography, Part IV. *Photographic Applications in Science Technology and Medicine.* May 1972.

Crossman, G. C. The Dispersion Staining Method for Selective Coloration of Tissue. *Stain Tech.* **24** 61-65 (1949).

Meyer-Arendt, J. The Photography of Colourless Microscopic Structures by the Colour Schlieren Method. *Photographie und Forschung* **5** 121-125 (1952)

Mitchison, J. M. and Swann, M. M1, Measurements on Sea-urchin Eggs with the Interference Microscope. *Quart. Jour. Microscop Sci.* **49** 381-389 (1953)

Saylor, C. P., Brice, A. T. and Zernicke, F. Color Phase Contrast Microscopy, Requirements and Applications. *Jour. Opt. Soc. Amer.* **40** 329-334 (1950)

Photomicrography

Barnard, J. E. and Welch, F. V. *Practical Photomicrography* Longmans, Green and Co. New York (out of print).

Gage, Simon Henry. *The Microscope. Seventeenth Edition.* Comstock Publishing Company (1941).

Corrington, J. D. *Working With The Microscope.* Whittlesey House (1941).

Shillaber, C. P. *Photomicrography in Theory and Practice.* John Wiley, New York (1944).

Michel, Kurt. *Die Wissenschaftliche Und Angewandte Photographie Wien.* Springer-Verlag (1957).

Schenk, Kistler, Bradley *Photomicrography.* Chapman and Hall Ltd. London (1962).

Clark, G. L. *The Encylopedia of Microscopy.* Rheinhold Pub. Corp. New York (1961).

Smith, R.F. Contrast Methods **In:** *The Microscopy of Living Tissue. Section IX, Tissue Culture Methods and Applications.* Academic Press New York (1973).

Grave, Eric V. *Discover the Invisible.* Prentice Hall New Jersey (1984).

Loveland, R. P. *Photomicrography, A comprehensive Treatise* 2 Vols, John Wiley, New York (1972).

Crossman, G. C. The Dispersion Staining Method for Selective Coloration of Tissue. *Stain Tech.* **24** 61-65 (1949).

Meyer-Arendt, J. The Photography of Colourless microscopic Structures by the Color Schlieren Method. *Photographie und Forschung* **5** 121-125 (1952).

Mitchison, J. M., and Swann. M. M., Measurements on Sea-urchin Eggs with the Interference Microscope. *Quart. Jour. Microscop. Sci.* **49** 381-389 (1953).

Saylor, C. P. Brice, A. T. and Zernicke, F. Color Phase-contrast Microscopy: Requirements and Applications. *Jour. Opt. Soc. Amer..* **40** 329-334 (1950).

Smith, R. F. The Use of Polarized Light in the Photography of Unstained Histological Radio Autographs of Plant Tissue. *Jour. Biological Photo. Assn.* **22** No. 1 (1954).

Smith, R. F. Differential Optical Staining of Colorless Living Organisms in Macro-Photography. *Jour. Biological Photo. Assn.* **23** Nos. 2 and 3. (1955)

Smith, R. F. A Variable Output Electronic Point Light Source for Photomicrography. *Jour. Biol. Photo. Assn.* **24** No. 2. (1957).

Smith, R. F. Behind the Lens Metering, The complete 35mm Photographic System. *Photographic Applications in Science Technology and Medicine.* **2** No. 7 (1968).

Smith, R. F. Color Contrast Methods in Microscopy and Photomicrography. *Photographic Applications in Science Technology and Medicine*. Part I, Vol.4, No. 17 (1970)

Smith, R.F. Color Contrast Methods in Microscopy and Photomicrography. *Photographic Applications in Science and Medicine*. Part II Vol. 5 No. 19 (1970).

Smith, R. F. Color contrast Methods in Microscopy and Photomicro graphy. *Photographic Applications in Science Technology and Medicine*. Part III. Vol.6 No.3 (1971).

Smith, R. F. Color Contrast Methods in Microscopy and Photo micrography, Part IV. *Photographic Applications in Science Technology and Medicine* (1972)

Smith, R. F. Micro Replication. Photographic Applications in Science. *Technology and Medicine*. 2 No.5 (1967).

Smith, R. F. The Microscopist's Microscope. *Functional Photography* 20 No.5 (1985)

Smith, R. F. Extending the Range of Optical Systems Photographic ally. *American Laboratory* 13 No.4 (1981)

Smith, R. F. Fiber Optics in Microscopy. *Functional Photography* (1986)

Smith, R. F. A Need for In-Depth Courses in Microscopy and Photomicrography. *Journal of Histotechnology* 11 No.3 (1988)

Smith, R. F. Preparation of Contact Microradiographs. *Industrial Photography*. (1989)

Smith, R. F. A Tribute to the Four Horsemen of Microscopy. *Functional Photography*. (1987)

olor Plate I

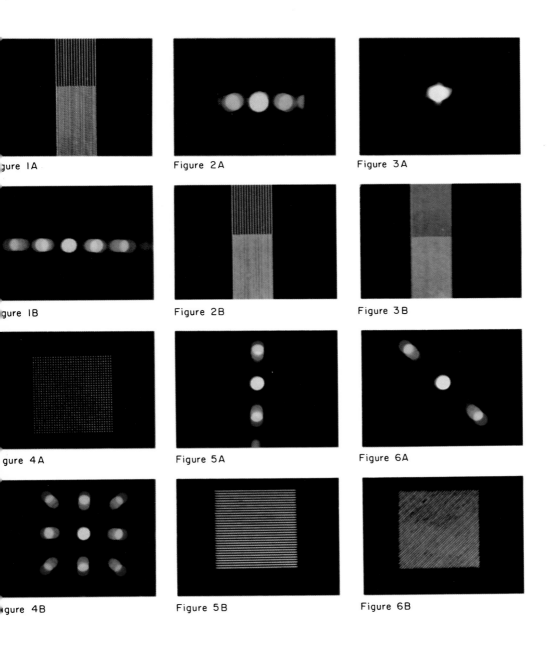

gure IA

Figure 2A

Figure 3A

gure IB

Figure 2B

Figure 3B

gure 4A

Figure 5A

Figure 6A

gure 4B

Figure 5B

Figure 6B

Color Plate II

RADIOLARIAN
TRANSMITTED LIGHT

RADIOLARIAN
TRANSMITTED LIGHT

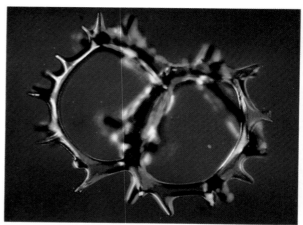

SURFACE OF METEORITE
REFLECTED LIGHT

COLOR FAULTS
ABBE CONDENSER

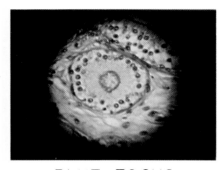

BLUE FOCUS

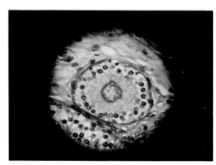

RED FOCUS

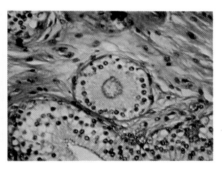

NO FILTRATION

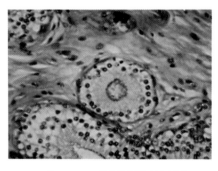

1mm DIDYMIUM

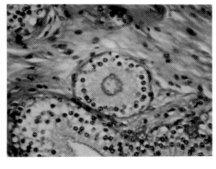

2 mm DIDYMIUM
OVER CORRECTION

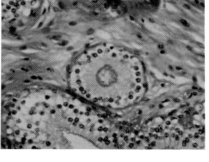

VOLTAGE TOO
LOW

Color Plate IV

HOW FILTERS CONTROL THE REPRODUCTION OF COLORS WHE
PHOTOGRAPHED WITH BLACK AND WHITE PANCHROMATIC FILM

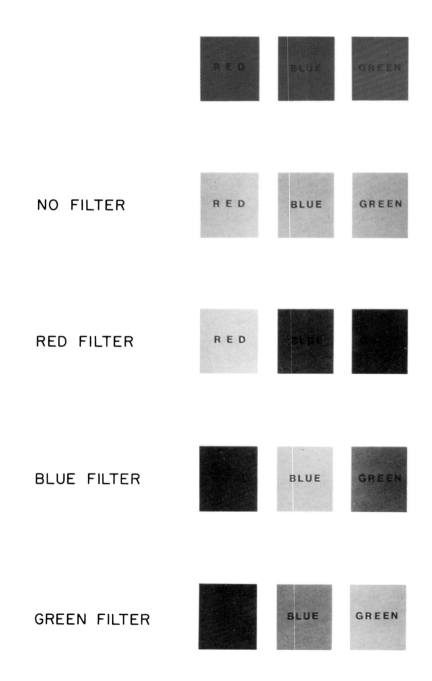

Color Plate V

THE ADDITIVE PRIMARY COLORS

RED FILTER
WRATTEN # 29

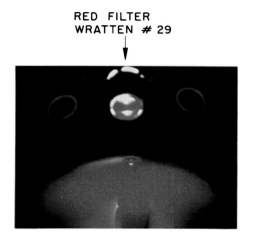

RED # 29   GREEN # 61

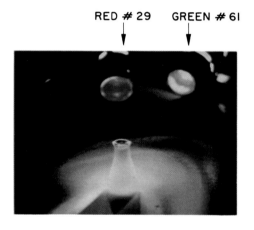

BLUE # 47   RED # 29   GREEN # 61

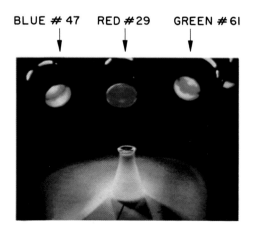

Color Plate VI - Demonstrating some of the staining possibilities possible using the additive light system.

A Shows a color diagram of the filtration. In this case it demonstrates below the stage illumination using a conventional microscope. Under these conditions the substage condenser must be removed to allow room for the fibers. Background illumination is handled in the same way as with the inverted microscope.

B The soda straws with the tri-color filters as explained previously

C Tri-color illumination with the soda straw filters.

D Specimen: *Clonorchis Sinensis*, fixed and stained 5X
Light source: Microscope illuminator brightfield.

E Specimen: *Clonorchis Sinensis* 5X
Light source: Single needle at 70°, filter. Pale blue gel in microscope illuminator at reduced voltage. Notice accent of internal details.

F Specimen: *Clonorchis Senesis* 5X
Light source: Single needle at 80° no filter, no background illumination. This image was made to demonstrate the selectivity of the system.

G Specimen: Young Starfish 10X
Illumination: Three needle fiber optics at 45° no filtration. no background illumination.

H Specimen: Young starfish 10X
Illumination: Three needle fiber optics at 45° Red, Blue, and Green filtration. No background illumination.

I Specimen: Young starfish 10X
Illumination: Three needle fiber optics at 45° Red, Blue, Green filtration. Magenta gel in background illuminator at reduced voltage.

J Specimen—Sugar crystals. Wet preparation 20X.
Illumination: Three needles Red, Green, and Blue filtration at 80°. No background illumination.

K Specimen: Pulverized glass particles 100X.
Illumination: Three needles Red, Green, and Blue filtration at 90°. No background illumination.

L Specimen: Fossil Radiolarian 100X
Illumination: Two needle fiber optics Red, and Green at 90°. Green gel in microscope illuminator at reduced voltage.

Color Plate VI

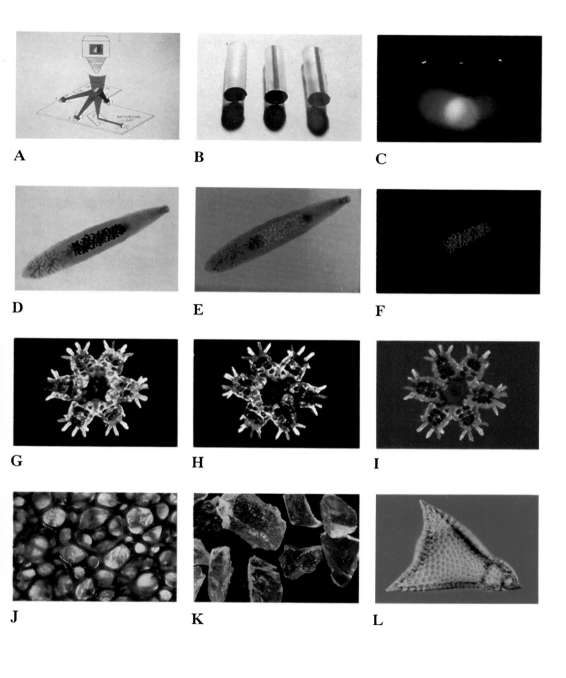

A

B

C

D

E

F

G

H

I

J

K

L

Index

P

parallax method, 89
phase contrast, 3, 42, 45, 56, 59-61, 63, 65, 67, 69, 103-104, 111, 134
photomicrography, 5, 23, 42, 63, 77, 79, 81, 83, 85, 87, 89-91, 93, 95, 97-99, 101-103, 105, 111, 113, 130, 133-135
photomultipliers, 102
polarizer, 63, 65

Q

Q-tips, 71

R

refraction, 3, 18, 28, 33, 35
refractive index, 1, 3, 28-29, 38, 49, 53, 59, 67

S

silicon photodiodes, 102
specimen plane, 14
spherical aberration, 42-43, 54, 56
stage, 9, 12, 16, 20, 23, 25, 27, 59-60, 85, 89-90, 115, 117, 136

T

technical pan, 98, 104-105
technidol, 104-106
troubleshooting, 71, 73, 75

W

Wollaston prisms, 63, 65